Design and Analysis of Simulation Experiments

Mathematics and Its Applications

Managing Editor:

M. HAZEWINKEL

Centre for Mathematics and Computer Science, Amsterdam, The Netherlands

Volume 339

Design and Analysis of Simulation Experiments

by

Sergei M. Ermakov
and
Viatcheslav B. Melas
Department of Mathematics and Mechanics,
St Petersburg State University,
St Petersburg, Russia

KLUWER ACADEMIC PUBLISHERS
DORDRECHT / BOSTON / LONDON

A C.I.P. Catalogue record for this book is available from the Library of Congress.

ISBN 978-90-481-4604-8

Published by Kluwer Academic Publishers,
P.O. Box 17, 3300 AA Dordrecht, The Netherlands.

Kluwer Academic Publishers incorporates
the publishing programmes of
D. Reidel, Martinus Nijhoff, Dr W. Junk and MTP Press.

Sold and distributed in the U.S.A. and Canada
by Kluwer Academic Publishers,
101 Philip Drive, Norwell, MA 02061, U.S.A.

In all other countries, sold and distributed
by Kluwer Academic Publishers Group,
P.O. Box 322, 3300 AH Dordrecht, The Netherlands.

This new and original work is based on the Russian book
С.М.Ермаков, В.Б.Мелас.
Математический эксперимент с моделями сложных стохастических систем. СПб. Изд-во С.-Петербургского ун-та, 1993.

Printed on acid-free paper

Contents

PREFACE

This book is devoted to a new branch of experimental design theory which is called simulation experimental design. The authors consider it necessary to start with a short introductory part in which some facts concerning random number generators and simple tools distribution simulation are presented, partly without proofs. This introduction is the content of the first chapter. Its first section could be titled "why can the multiplication generator be used in simulation ?". Asymptotical properties of this generator and the limit theorem first discussed in [29] can be the basis for applying the multiplication generator in simulation and explains the necessity of usage of the generator with a great number of digits. The authors by no means deny the role of numerous papers devoted to investigation and testing of various modification of the generator, but it seemed important to formulate a consistent point of view on the matter if only in the simplest case. Sections 2 and 3 of Chapter 1 are not original and are given for the convenience of begining readers.

One of the main topics of the book is the elaboration of the splitting and Russian roulette approach to the design of simulation experiments. Some advantages of this approach were stressed twenty years ago [29] in comparison with the better known importance sampling technique. In this book, mainly by efforts of the second author, this approach was further developed into a mathematical theory. The other topic of the book consists of the usage of regression models with a systematical error (bias) and in the development of models to be robust in relation to this error. This idea also goes back to [29] , but here it obtains a more complete form.

Note that throughout the book variance is denoted by the symbol $\mathcal{D}$ (dispersion) according to Russian custom.

The authors are indebted to Mr. A.Sokolov for the help in the translation of the book into English and to Ms. I.Smirnova for the printing of the typescript.

This work was partly supported by ISF Grant Number NVZ 000.

St.Petersburg, March 1995

The Authors

CHAPTER 1. METHODS OF DISTRIBUTION AND PROCESS SIMULATION

1. The arithmetic modeling of randomness

In this chapter, we will consider some aspects of the problem of randomness simulation. More exactly, we will try to find out how to obtain the same results by deterministic means as when observing the phenomena, described by probability theory.

A "random-number generator" certainly is a part of the software used in complex probabilistic systems simulation. If this generator is a program, then it is supposed that the use of this program will allow one to obtain (in computer representation) a real number α from the interval $[0,1]$, so that a sequence of numbers $\alpha_1, \alpha_2, \ldots$, formed by this can be considered as a sequence of independent realizations of a random value having a uniform distribution at the interval $[0,1]$.

It is known that for computer simulation an often-used generator has to be simple (fast). At the same time, one can guess that a rather simple algorithm cannot yield a sequence similar to a random one. A.N. Kolmogorov and his successors [102,17,63] gave the exact definition of the problem and showed results regarding the completeness of algorithms producing random sequences. It turns out that such algorithms must be very (infinitely in limit) complex. This could have been an insurmountable barrier for the creation of a programmed random number generator. However, some simplification of the problem of the simulation of random values allows one to be content with a comparatively simple generator. Sequences formed by these generators are called pseudorandom sequences. One such generator which has become quite well known, is the Lehmer multiplicative generator. This generator forms a sequence of numbers according to the formulae

$$\begin{aligned} X_{n+1} &\equiv M x_n \,(\mathrm{mod}\ P), \\ \alpha_{n+1} &= x_{n+1}/P. \end{aligned} \tag{1.1}$$

Here x_n, M, P are positive integers ; the sign "$\equiv$"(comparison) means that x_{n+1} is the remainder on dividing Mx_n by the nearest whole power P. If, in a computer realization, a whole number is represented by no more than m binary digit bits, then $P = 2^m$, $M < 2^m$ is taken. The sequence has a period of which the lenght does not exceed 2^{m-2}. The maximal length is provided with a special selection of x_0 and M. The corresponding values of these numbers may be found in [29, 46].

The length of a period may be increased by using more general formulae for the value x_n, namely:

$$x_{n+1} = M_1 x_n + M_2 x_{n-1} + \ldots + M_k x_{n-k+1} + M_{k+1} (\mathrm{mod}\ P), \quad k \geq 1. \tag{1.2}$$

It is assumed that $k = 1$ (a mixed generator).

It is also advisable to use a combination of several generators, one of which can determine the operation order of the others. For example, if using five generators, then the first one forms a sequence $\alpha'_1, \alpha'_2, \ldots$, where $0 < \alpha'_i < 1$; the first two digits of each of these numbers will indicate the number of that generator of the remaining four generators, which will calculate the element sought from the main sequence.

In most papers on generator problems, the use of pseudorandom sequences is justified by the results of their verification by statistical tests. However, some authors correctly point out that, instead of statistical tests, one can easily use applied problems, solved by the Monte Carlo method. The problem of testing pseudorandom sequences is widely covered, e.g. in [36,46] where all the necessary information can be found. Such verification is, without doubt, interesting. But it must be admitted that the problem of test system completeness has not been exactly formulated. Indeed, it is always easy to see why a sequence should be "rejected", whereas it is unclear which characteristic shows that it is good, i.e. genuinely random, uniformly distributed, etc.

We shall not consider types of statistical tests and analyze the results of their application as there are papers on these points available. Instead, let us dwell on the number-theoretic approaches.

It is common knowledge that the Law of Large Numbers and the Central Limit Theorem are fundamental for probability theory. Even their simplest form can provide the basis for many simulation algorithms.

Let us recall the strong Law of Large Numbers for independent equally distributed random values.

Let $\xi_1, \xi_2, \ldots, \xi_N$ be independent realizations of a random value ξ and let its mean value $E\xi$ exist. Then, with probability 1,

$$\lim_{N\to\infty} \frac{1}{N} \sum_{i=1}^{N} \xi_i = E\xi. \tag{1.3}$$

If $E\xi^2 < +\infty$, then the following simplest version of the Central Limit Theorem is valid

$$\lim_{N\to\infty} P\left(\left|\frac{1}{N}\sum \xi_i - E\xi\right| \frac{\sqrt{N}}{\sigma} < \gamma\right) = \int_{-\gamma}^{\gamma} \frac{1}{\sqrt{2\pi}} e^{-\frac{1}{2}u^2} du, \tag{1.4}$$

where

$$\sigma^2 = E\xi^2 - (E\xi)^2.$$

Suppose now that we consider only such random values whose realizations can be calculated on the basis of a finite or countable number of ideal coin-tossing (which will be considered in detail later). Random values allowing for the simulation by means of such an algorithm are called 'constructively given'.

If we have a true generator of random numbers (independent, uniformly distributed), then the simulated value ξ can be expressed in the form

$$\xi = g(\alpha_1, \ldots, \alpha_s), \tag{1.5}$$

where $\alpha_1, \ldots, \alpha_s$ are independent and uniformly distributed, and s, being infinite, is finite in actual calculations but can be a large number. Note that for random numbers of such a type, equality (1.3) has the form

$$\lim_{N\to\infty} \frac{1}{N} \sum_{i=1}^{N} g(\alpha_{is+1}, \ldots, \alpha_{(i+1)s}) = \int_0^1 dx_1 \ldots \int_0^1 dx_s\, g(x_1 \ldots x_s). \tag{1.6}$$

Let there be a sequence (nonrandom) of real numbers at the segment $[0,1]$

$$\mathcal{E} = \varepsilon_1, \varepsilon_2, \ldots, \varepsilon_N \ldots$$

It is evident that if equalities similar to (1.6) are fulfilled, then it is possible to use the numbers from $\mathcal{E}$ when simulating randomness. Equality (1.6) is accomplished with probability 1. That condition is not justified from the deterministic point of view. It is necessary that the equality

$$\lim_{N\to\infty} \frac{1}{N} \sum_{i=1}^{N} h(\varepsilon_{is+1}, \ldots, \varepsilon_{(i+1)s}) = \int_0^1 dx_1 \ldots \int_0^1 dx_s\, h(x_1 \ldots x_s) \tag{1.7}$$

holds for any integrable function in s variables $h(x_1 \ldots x_s)$ and for any s. If the integrability is the Lebesgue one, then (1.7) cannot be valid for all h ($\mathcal{E}$ has the Lebesgue measure 0) and it is necessary to restrict the class of functions at least to a class of functions integrable in the Riemann sense.

After that, we can see that the sequences, for which equality (1.7) is valid, exist. They have been studied in number theory. For $s = 1$, the corresponding sequence is called 'the uniformly distributed sequence in the Weil sense' [103]. Such a sequence is exemplified by

$$\varepsilon_{n+1} = \{n\beta\}, \tag{1.8}$$

where β is any irrational number, $\{a\}$ is a fractional part of the number a. A sequence for which (1.7) holds for all s has been constructed by I.M. Korobov [64]. Such sequences are called "completely uniformly distributed".

Let us return to the standard generator (1.1) and note the following

1. Calculations by formulae (1.1) are carried out with finite register (binary) lengths. With $P = 2^m$, we manipulate in binary arithmetic with numbers having a finite number of binary digits.

2. Taking into account the first remark, formulae (1.1) may be written in the form

$$\alpha_{n+1} = \{M\alpha_n\}. \tag{1.9}$$

For the case when α_n may be any real number (there is no finiteness demand for the number of bits in a computer representation α_n), sequence (1.9) has been fully considered in number theory. It is usually written in the form

$$\varepsilon_{n+1} = \{M^n \varepsilon_1\} \tag{1.10}$$

and is called "the sequence of fractional parts of the exponential function".

It is known [42] that sequence (1.10) is completely uniformly distributed, asymptotically with respect to M. Theorem 1.1 takes place for any integrable function s in $h(x_1, \dots, x_s)$ variables in respect of Riemann, any s, and for almost all ε_1 from $(0,1)$

$$\lim_{M\to\infty} \lim_{N\to\infty} \frac{1}{N} \sum_{j=0}^{N-1} h(\varepsilon_{js+1}, \dots, \varepsilon_{(j+1)s}) = \int_0^1 dx_1 \dots \int_0^1 dx_s\, h(x_1, \dots, x_s) \qquad (1.11)$$

(an analogue of the Law of Large Numbers). Moreover, in [28] it is shown that an analogue of the Central Limit Theorem exists, namely,

Theorem 1.1. *If $h(x_1, \dots, x_s)$ is any square integrable in a unit hypercube $\mathcal{D}_s$ function ($\mathcal{D}_s = \{0 \le x_i \le 1,\ i = 1, \dots, s\}$), then the equality*

$$\lim_{M\to\infty} \lim_{N\to\infty} \text{mes} \left\{ \varepsilon_1 : \left| \frac{1}{N} \sum_{j=0}^{N-1} h(\varepsilon_{js+1}, \dots, \varepsilon_{(j+1)s}) - J[h] \right| \frac{\sqrt{N}}{\sigma_h} < \gamma \right\} = \\ = \int_{-\gamma}^{\gamma} \frac{1}{\sqrt{2\pi}} e^{-\frac{1}{2}u^2} du, \qquad (1.12)$$

is valid, $J[h] = \int_0^1 dx_1 \dots \int_0^1 dx_s\, h(x_1, \dots, x_s)$ and

$$\sigma_h^2 = \int_0^1 dx_1 \dots \int_0^1 dx_s\, h^2(x_1, \dots, x_s) - (J[h])^2. \qquad (1.13)$$

Both results may be obtained (see [29]) from Xiu's lemma that says

Lemma 1.1. *Let $h(x_1, \dots, x_s)$ be integrable with respect to Riemann in $\mathcal{D}_s$ and $\varphi_M(x) = h(x, \{Mx\}, \dots, \{M^{s-1}x\})$, then*

$$\lim_{M\to\infty} \int_0^1 \varphi_M(x)\, dx = \int_0^1 dx_1 \dots \int_0^1 dx_s\, h(x_1, \dots, x_s). \qquad (1.14)$$

We shall not prove the lemma and the theorems because their proofs can be found in [29]. Note that $h(\varepsilon_{js+1}, \dots, \varepsilon_{(j+1)s}) = h(\{M^{js}\varepsilon_1\}, \dots, \{M^{(j+1)s-1}\varepsilon_1\}) = \varphi_M(\{M^{js}\varepsilon_1\})$ and $\int_0^1 \varphi_M(\{M^{js}x\})dx = \int_0^1 \varphi_M(x)\, dx$. Hence the assertion of Theorem 1.1 is easily obtained. The proof of Theorem 1.2 is more complicated.

It is clear that the given results may be used for justifiying the use of the Monte Carlo method for calculating with an infinite (at least with a very great) number of significant digits.

Direct calculations are, however, carried out with a finite number of bits. As we have already mentioned, there are a lot of papers where the generator (1.1) has been investigated theoretically and by means of tests. Further, we shall see to what extent the approach based on Xiu's Lemma can be applied to the case of a finite number of bits.

Let us assume that the calculations are performed with m binary bits. Then an analogue of the integral in equality (1.14) should be the sum

$$S(m, M) = \frac{1}{2^m} \sum_{j=0}^{2^m-1} \varphi_M\left(\frac{j}{2^m}\right), \tag{1.15}$$

where

$$\varphi_M\left(\frac{j}{2^m}\right) = h\left(\frac{j}{2^m}, \left\{\frac{jM}{2^m}\right\}, \dots, \left\{\frac{jM^{s-1}}{2^m}\right\}\right).$$

Since h, by supposition, is Riemann integrable, then for each fixed M, $\varphi_M(x)$ is also Riemann integrable at the segment $[0,1]$ provided that

$$\lim_{m\to\infty} S(m, M) = \int_0^1 \varphi_M(x)\, dx. \tag{1.16}$$

From the latter equality, with some necessary modifications, one may obtain an analogue of Theorem 1.1 pertaining to calculations with m bits. To that end, it is necessary to introduce a measure μ_m to the finite set T_m, $T_m = \{j/2^m;\ j = 0, 1, \dots, 2^m - 1\}$ of the points such that

$$\mu_m\left\{\frac{j}{2^m}\right\} = \frac{1}{2^m}, \quad j = 0, 1, \dots, 2^m - 1.$$

If we select $\varepsilon_1^{(m)} \in T_m$ and consider the sequence

$$\varepsilon_{n+1}^{(m)} = \left\{M\varepsilon_n^{(m)}\right\}, \tag{1.17}$$

then the following theorem is valid.

Theorem 1.1a. *For every Riemann integrable function $h(x_1, \dots, x_s)$ in s variables, every fixed s and every $\varepsilon > 0$, the equality*

$$\lim_{N\to\infty} \lim_{M\to\infty} \lim_{m\to\infty} \mu_m \left\{ \varepsilon_1^{(m)} : \left| \frac{1}{N} \sum_{j=0}^{N-1} h\left(\varepsilon_{js+1}^{(m)}, \dots, \varepsilon_{(j+1)s}^{(m)}\right) - \int_{D_s} h(X)\, dX \right| > \varepsilon \right\} = 0. \tag{1.18}$$

holds.

The analogue of Theorem 1.2 may also be obtained for finite m but we shall not give it here.

Note that, according to the received qualitative result, it is advisable to use a "generalized" multiplicative generator given by the formulae

$$\begin{aligned} x_{n+1} &\equiv M x_n \,(\operatorname{mod} 2^{m_1}), \\ \alpha_{n+1} &= \lfloor 2^{m-m_1} x_{n+1} \rfloor / \, 2^m, \end{aligned} \tag{1.19}$$

where $m_1 > m$ and the brackets mean the whole part of a number. Formulae (1.19) mean that we can obtain a pseudorandom number by carrying out calculations with a larger number of bits (the first formula of (1.19) and then keep only the first m bits.

There are works in which some advantages of the generator (1.19) are discussed. In [37], tests on integral calculations by means of the generator (1.19) are described. In [61] it is shown that the complexity of predictional algorithms for (1.19) with the fixed m increases exponentially if m_1 increases. This latter result brings the number-theoretical approach closer to the approach based on the notion of the complexity of algorithms.

Let us make some additional remarks.

Remark 1. The equality (1.12) is relative to an explicit function f. From this equality we can conclude that ε_1 exists, and that, for any finite set of functions f_j, $j = 1, \ldots, L$ for sufficiently large M, N (in a strictly determinate sense), the inequality

$$\left| \frac{1}{N} \sum_{j=0}^{N-1} h\left(\varepsilon_{js+1}^{(m)}, \ldots, \varepsilon_{(j+1)s}^{(m)}\right) - \int_{\mathcal{D}_s} h(X)dX \right| \leq \frac{y\sigma_f}{\sqrt{N}} \tag{1.20}$$

will be carried out. Here, y may be large for large L. An estimate for y may be found in [29]. In this connection, M.V. Levit work [72] is of interest, where for a whole class of functions, parameters of the random number generator are such that (1.20) is fulfilled.

Remark 2. By theoretical reasoning connected with the estimate's convergence to a mean value expressing integral M, it is, strictly speaking, necessary to consider a set of generators (1.17) with increasing m and M.

The stated results provide a theoretical basis for one of the ways of arithmetic simulation of a uniformly distributed random value. The literature on this topic is quite extensive. We shall refer to [71] which is of principal importance to appreciate the heart of the matter.

Another random generator, which has the properties of the asymptotically complete uniform distribution and is independent, is the Tauswort [101] generator. It has the following structure. Let $c_1, \ldots, c_n$ be a given set of zeros and units, $c_1 = 1$. Numbers a_k, $k = 1, 2, \ldots$, are defined by the recurrence relation

$$a_k \equiv c_1 a_{k-1} + c_2 a_{k-2} + \ldots + c_n a_{k-n} \ (\operatorname{mod}\ 2). \tag{1.21}$$

A vector $a_m, a_{m-1}, \ldots, a_{m-n+1}$ is repeated in no more than $p = 2^n - 1$ steps. The sequence has a period. This period is maximal if and only if the polynomial

$$f(x) = 1 + c_1 x + \ldots + c_n x^n$$

is primitive over field $GF(2)$. Primitive polynomials over $GF(2)$ have been scrutinized and tabulated data are available [101].

Let us set up a binary number

$$y_k = 0, \quad a_{qk-1}a_{qk-2}\dots a_{qk-L} \text{ (a fraction in the binary representation 2)}, \tag{1.22}$$

where $L \leq n$. This fraction is of L binary terms of the sequence (1.21), $q \geq L$ and $2^n - 1$, q are supposed to be relative primes.

Sequence y_k possesses the necessary properties too. For $L \to \infty$, $n \to \infty$, the sequence asymptotically has completely uniform distribution, and for Riemann integrable functions f in $\mathcal{D}_s$, the analogue of Theorem 1.2 holds where ε_1 acts as y_0 [26].

Besides, it is also important, that the bit operations on separate bits of numbers are easily realized by computer software, and that very fast random number generators of the type (1.1) may be devised.

Thus, to justify the applicability of a pseudorandom number generator, we did not use ideas concerning the relation between the nature of randomness and the complexity approach. However, owing to this simplification of the problem, we have to narrow down the class of random values under consideration and study only the random values of the $f(\alpha_1, \alpha_2, \dots, \alpha_s)$ type, where f is Riemann integrable. But, in fact, it is necessary that f belongs to a narrower class (e.g., meets the Lipschitz condition). Otherwise serious errors connected with the finiteness of values M and m can arise, and the use of Theorems 1 and 2 will not be justified. Actually, the finiteness of M and m leads to problems which should be considered separately.

As was shown above, some results of theoretical investigations exist that enable us to guarantee the fulfillment of many required generator properties without tests; at least in the problems connected with the Law of Large Numbers and the Central Limit Theorem. On the other hand, tests prove the good quality of the multiplicative generator and its modifications [36] and, with finite m and M, they are an important supplementary device for the evaluation of the sequence quality. From that given above, it follows that the statement "a random number generator is not suitable since it is not random" is based on "everyday life" understanding of randomness. Attempts to "improve" the sequence by an "absolutely" random choice of the sequence's initial numbers provide nothing but the possibility of an error in calculations (the possibility of degeneracy of a sequence).

Along with this, under insufficiently large m and M the terms of the sequence generated by multiplicative generators are considerably dependent on each other. For this reason, as was pointed out, their intermixing by means of an other generator proves desirable, as was indicated above. Of course, if the unpredicted behavior of an obtained sequence is required from a random number generator, then a multiplicative generator is unsatisfactory, since its values are completely determined by the two parameters M and m. If it is known that a multiplicative generator is used, then m and M can be regenerated by a few observations, and, therefore, the whole sequence of numbers can be predicted. This undesirable, from the viewpoint of some users, property proves to be very useful in calculating. It enables one to obtain identical results when iterating calculations (to carry out an iterated count).

Besides the pseudorandom numbers, the so-called quasirandom numbers are sometimes used for simulating. The Law of Large Numbers is valid for sequences formed by these numbers, whereas the Limit Theorem in the form of (1.4) is not. For these sequences at a class, narrower than the Riemann integrable functions class, an integrating error decreases more rapidly than $N^{-1/2}$. Most quasirandom sequences may be statistically treated as the sequences of uniformly distributed and yet dependent values. They should be used carefully for the imitation of random processes since the second moments values of the imitated process can differ from those required.

Note that in the recent book of Niderreiter [88] many properties of the multiplicative generator and its generalizations are studied and can be of interest for the reader.

2. Simulation of discrete distributions

Many complex systems are described by a Markov chain with a finite or countable number of states. The concept of such a chain is introduced by the description of its simulation algorithm. As this algorithm includes as its component algorithms of the simulation of discrete distributions, we have to dwell briefly on the methods used for such a simulation.

Let a random value ξ take values $x_1, x_2, \dots$ with probability $p_1, p_2, \dots$ respectively, and $p_i > 0, \;\; i = 1, 2, \dots, \;\; \sum p_i = 1$. Then ξ is said to have a discrete distribution

$$\begin{pmatrix} x_1, x_2, \dots, x_n, \dots \\ p_1, p_2, \dots, p_n, \dots \end{pmatrix}. \tag{2.1}$$

Since x_i is uniquely determined by its number i, then instead of (2.1), the following distribution is always possible

$$\begin{pmatrix} 1, & 2, & \dots, & n, & \dots \\ p_1, & p_2, & \dots, & p_n, & \dots \end{pmatrix}. \tag{2.2}$$

In Section 1 the arithmetic basis for the simulation of uniformly distributed independent random values has been discussed. Further, we assume that a computer with which the calculations are carried out has in its software such a command (macrocommand) that, when executing that command, a number α possessing the necessary properties of an independent random value, which is uniformly distributed at the segment $[0, 1]$, appears in a given cell of the memory (register).

Let us call the numbers obtained in such a way "random numbers" and denote them by α with different indexes.

Let the (discrete) probability distribution (2.2) be given. The objective is to indicate an algorithm that enable us to calculate by some system of random numbers α, the realization of a random value assuming n with probability $p_n, \;\; n = 1, 2, \dots$.

Algorithm 1. Is based on the following (inversion formula).

Let a random value ξ take n_i with probability p_i, $i = 1, 2, \dots, N$ and α be a uniformly distributed random value at $[0, 1]$. Then a number α, for which an equality

$$\sum_{j=1}^{n-1} p_j \le \alpha < \sum_{j=n}^{N} p_j,$$

is valid, is a realization of the random value ξ.

In Figure 1 one can see the procedure of a simple proof.

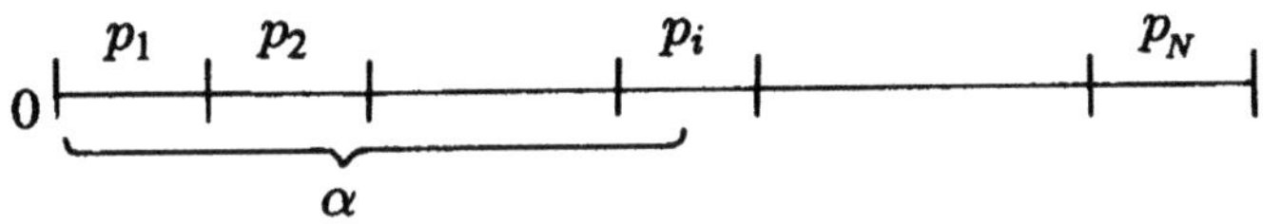

FIG. 1

Short segments of length $p_1, p_2, \dots$ are successively plotted on $[0, 1]$. From the definition of the uniformly distributed random quantity, it follows that the probability of a random number α hitting a segment of length p_i is just equal to p_i.

Algorithm 1 consists in successively subtracting $p_1, p_2, \dots$ from α until a negative number is obtained. It is evident that the number of calculations depends on the numbering of p_i. The first numbers must have the greatest probabilities.

When simulating, various methods for the search in tables (e.g., bisection, by which a table is successively divided into two parts) can be applied. For large N, if the values p_i differ little (or for infinite N if a series $\sum p_i$ converges slowly), obtaining a realization ξ can take a computer a long time. That is why various algorithms of simulation are used in different cases. We shall show some of them to give a more complete understanding of the up-to-date simulation techniques.

Algorithm 2. (Rejection method in the discrete case).
Let $p = (p_1, \dots, p_N)$ and $q = (q_1, \dots, q_N)$ be two probability distributions and a constant c be such that $c \ge 1$ and $p_i \le cq_i$, $i = 1, 2, \dots, N$. Using the next algorithm, we can obtain a random quantity ξ which has a distribution p:

1. we get x, which has a distribution q; we have a uniformly distributed random value α;
2. check the inequality $\alpha c q_x \le p_x$;
3. if the inequality is valid, then $\xi = x$, otherwise go to 1).

The distribution q is called the majorizing distribution. The closer q is to p, the more efficient this method becomes. However, the simulation algorithm of q should be much simpler than the simulation algorithm of p. Otherwise, the application of this method makes no sense.

For separate distributions (particularly for infinite N), specially selected majorants exist which allow us to construct algorithms for efficient simulating. These majorants, together with a simple justification of the algorithm, can be found in the extensive monographs [3,25].

The next example which may be useful for diminishing the working time when simulating distributions, is the representation of a distribution as a combination of two or more distributions. If the following representation exists (a combination of two distributions)

$$\begin{pmatrix} 1, & 2, & \ldots, & N \\ p_1, & p_2, & \ldots, & p_N \end{pmatrix} = \\ = q_1 \begin{pmatrix} 1, & 2, & \ldots, & N \\ p_1^{(1)}, & p_2^{(1)}, & \ldots, & p_N^{(1)} \end{pmatrix} + q_2 \begin{pmatrix} 1, & 2, & \ldots, & N \\ p_1^{(2)}, & p_2^{(2)}, & \ldots, & p_N^{(2)} \end{pmatrix}, \tag{2.3}$$

then we can use

Algorithm 3.

1. simulate a random value i with distribution $\begin{pmatrix} 1 & 2 \\ q_1 & q_2 \end{pmatrix}$;
2. simulate (e.g., by either Algorithm 1 or Algorithm 2) the distibution

$$\begin{pmatrix} 1, & 2, & \ldots, & N \\ p_1^{(i)}, & p_2^{(i)}, & \ldots, & p_N^{(i)} \end{pmatrix}$$

for a random value ξ_i; the value arrived at is the realization of a random value with distribution (2.3).

The generalization of the algorithm for the general case of representation (2.3) is evident in the form

$$p_j = \sum_{i=1}^{L} q_i p_j^{(i)}, \quad j = 1, \ldots, N, \quad q_i > 0, \quad \sum_{i=1}^{L} q_i = 1.$$

In step 1 of Algorithm 3 i, having a distribution

$$\begin{pmatrix} 1, & 2, & \ldots, & L \\ q_1, & q_2 & \ldots, & q_L \end{pmatrix},$$

ought to be simulated.

Example. Sometimes it is useful to have the representation in the form of (2.3) wherein $p_{j_1}^{(1)} = p_{j_2}^{(1)} = \ldots = p_{j_s}^{(1)} = 1/s$, $p_{j_s+1}^{(1)} = \ldots = p_{j_N}^{(1)} = 0$, $j_1 < j_2 < \ldots < j_s$, $1 \le s \le N$ i.e., to choose a discrete uniform component. If it turns out that q_2 is small, then a considerable advantage can be obtained, since a discrete distribution in the form

$$\begin{pmatrix} 1, & 2, & \ldots, & s \\ 1/s, & 1/s, & \ldots, & 1/s \end{pmatrix}$$

is simulated very simply by means of the following algorithm

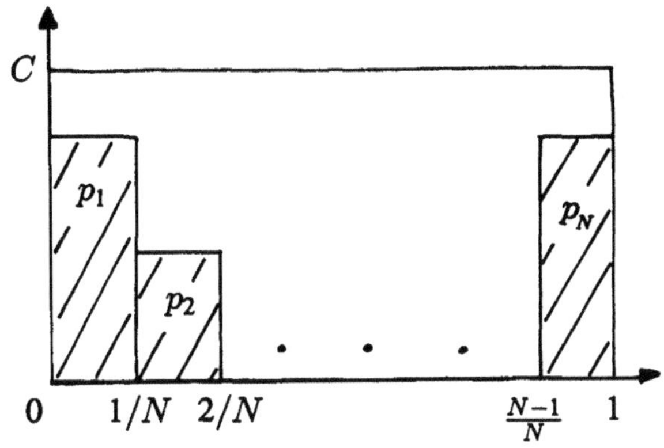

FIG.2

Algorithm 4.
1. get a uniformly distributed random number α;
2. assume $\xi = \lfloor s\alpha \rfloor$.

In binary ariphmetic, if $s = 2^m$, then step 2 in the algorithm is carried out very simply.

Consider Figure 2. A rectangle with number i has an area p_i and a lower base $1/N$ in length. A constant C is such that $C \geq \max_i p_i$. Choose a uniformly distributed α on $[0,1]$, α' – on $[0,C]$, and suppose $j = \lfloor N\alpha \rfloor$, j is a realization of a random value with distribution p if $\alpha' \leq p_j$. Otherwise, another pair α, α' is needed. This is a version of the selection method with a constant majorant.

One can construct a similar method wherein there are no unsuitable α and α' in the shown sense. The corresponding Algorithm 5 may have important advantages. It is based on Walker's theorem [25].

Theorem. *Every probabilistic vector* $(p_1, \ldots, p_N)$ *may be represented as an equiprobable mixture N of two-point distributions.*

Proof. Let $1\,[i = j]$ mean a function of integer-valued arguments i and j. When $i = j$, it equals unity, otherwise it equals zero. We shall demonstrate that there are N pairs of integers $(i_1, j_1), \ldots, (i_N, j_N)$ and that probability distribution $q_1, \ldots, q_N$ is such that the equalities

$$p_i = \frac{1}{N} \sum_{l=1}^{N} (q_l\, 1\,[i_l = 1] + (1 - q_l)\, 1\,[j_l = i]), \tag{2.4}$$

are fulfilled which will prove our theorem. The proof will be produced by induction. It is evident that when $N = 2$, the above numbers and the probability distribution exist. Assume that they exist when $N < n$, $n > 2$; then prove the existence for $N = n$. Choose the least p_i. It is clear that it does not exceed a quantity $1/N$, and we can assume $i_1 = \arg\min_i p_i$ and $q_1 = Np_{i_1}$. Then assume $j_1 = \arg\max_i p_i$. A pair of combinations is considered by this to be definite (equal to one of the addends of the sum in equality (2.4)).

Note that $(1-q_1)/N \le p_{j_1}$ since $p_{j_1} \ge 1/N$. Now eliminate the probability p_{i_1} in the initial distribution, replace p_{j_1} by $p_{j_1} - (1-q_1)/N$. The obtained probabilistic vector has a degree $N-1$ and the sum of its components is $(N-1)/N$. After the renormalization of this vector, the above assumption will be valid, which proves the theorem.

Since (2.4) has been obtained (the way of its construction is found in the proof of the theorem), then the following algorithm for simulating random value ξ takes place

Algorithm 5.
1. get uniformly distributed α and α';
2. assume $l = \lfloor \alpha N \rfloor$;
3. if $q_l < \alpha'_2$ then $\xi = i_l$, otherwise $\xi = j_l$.

The above-mentioned means of simulating discrete distributions may be combined in a different way. Of course, the question can arise of how the optimal algorithm for the simulation of the given distributions can be found among all possible algorithms.

The answer to this question depends on many factors and its solution cannot be simple. For example, it depends on the way of receiving uniformly distributed random numbers, the number of elementary steps that a given computer needs to execute one or another operation, the volume of rapid and extrarapid memory, the availability of parallel structures, etc.

In Knuth and Yao's paper [62] a feasible approach to the choice of the optimal simulation algorithm is considered. The problem is solved under the following assumptions:

1) we have a source of independent random bits taking values 0 or 1 with probability 1/2 (an imitator of flipping an ideal coin);

2) random bits are successively and independently generated. After that, the bits of an independent realization of a random quantity with a given distribution law are generated.

In distribution (2.2), let the probabilities p_j have the binary expansions

$$p_j = (0, \varepsilon_1^j \varepsilon_2^j \ldots \varepsilon_n^j \ldots)_2, \quad \varepsilon_k^j = 0, 1, \quad k = 1, 2, \ldots \tag{2.5}$$

A sequential appearance of independent bits $\alpha = (0, a_1 a_2, \ldots, a_n \ldots)_2$, $a_j = 0$ or 1 and their comparison with (2.5) enables us to construct a "tree", which illustrates a simulation process.

Let k_1 be the first value from k for which not all $\varepsilon_{k_1}^j$ are equal to zero, and $j_1^{(1)}, j_2^{(1)}, \ldots,$ $j_{s_1}^{(1)}$ are corresponding values j represented in (2.5). Similarly, k_l is the lth value from k for which there are corresponding numbers equal to unit $\varepsilon_{k_l}^j$, $j_1^{(1)}, \ldots, j_{s_l}^{(1)}$.

Our distribution may be given as a sum

$$\begin{pmatrix} 1, & 2, & \ldots, & N \\ p_1, & p_2, & \ldots, & p_N \end{pmatrix} = \begin{pmatrix} j_1^{(1)} & j_2^{(1)}, & \ldots, & j_{s_1}^{(1)} \\ 1/2^{k_1}, & 1/2^{k_1}, & \ldots, & 1/2^{k_1} \end{pmatrix} + \tag{2.6}$$

$$+ \begin{pmatrix} j_1^{(2)} & j_2^{(2)}, & \ldots, & j_{s_2}^{(2)} \\ 1/2^{k_2}, & 1/2^{k_2}, & \ldots, & 1/2^{k_2} \end{pmatrix} + \ldots + \begin{pmatrix} j_1^{(l)} & j_2^{(l)}, & \ldots, & j_{s_l}^{(l)} \\ 1/2^{k_l}, & 1/2^{k_l}, & \ldots, & 1/2^{k_l} \end{pmatrix} + \ldots$$

Under the corresponding normalization, (2.6) can be rearranged into (2.3) for the combination of discrete uniform distributions. However, in this case, a random number generator is structured in such a way that it is inconvenient to reduce the simulation process to the simulation of a combination (more exactly, the combination will be simulated in a special way).

We obtain $a_1, \dots, a_{k_1}$ (k_1 independent flippings of an ideal coin). Then a binary number $(0, a_1 \dots a_k)$ is a required realization of a random quantity, if it coincides with one of the numbers $j_1^{(1)}, \dots, j_{s_1}^{(1)}$. In another case, we go to another addend, and get $a_{k_1+1}, \dots, a_{k_2}$. If the number $(0, a_{k_1+1}, \dots, a_{k_2})$ coincides with one of $j_1^{(2)}, \dots, j_{s_2}^{(2)}$, then it is a required realization. In another case, we go to the next addend, and get $a_{k_2+1}, \dots, a_{k_3}$, etc.

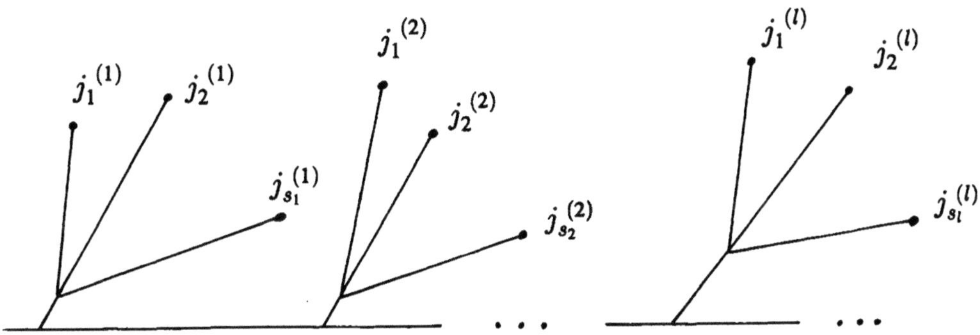

FIG. 3

Such a process can be visually represented in the form of a tree (Figure 3). Trees of this kind are called GDD trees (trees generating a discrete distribution). Other ways of constructing trees of that kind are possible. If assumptions 1 and 2 are fulfilled, a GDD tree corresponds to each distribution. Among these trees, one can choose a tree corresponding to the least mean number of flipping a coin. A simulation algorithm, connected with this tree, is called the optimal GDD algorithm. The theory of a GDD tree can be found in [62].

We are paying much attention to the methods of the simulation of discrete distributions, because they are very important for further consideration of the simulation of Markov chains with a finite or infinite number of states.

Besides, for the simulation of complex systems, the simulation of random values, given by their distribution density, and random vector simulation are important. Many papers devoted to these problems have been published recently. Simulation algorithms for the most important distributions (exponential, normal, and some others) have been developed and considerably improved. The methods of distribution simulation can be found in papers by Russian scholars [36, 99]. An extensive and encyclopedic monograph [24] contains a present-day view on the problem. This book is necessary for designers of distribution simulation algorithms. Its main ideas will also be interesting for users, because they may help into critically evaluating the algorithms used.

Let us outline these ideas. They allow one to obtain an economical algorithm for distribution simulation when such a simulation is occasionally needed. If constant use of the algorithm is required, it should be performed more thoroughly, taking into account all the experience accumulated in this area. It should be done by experts.

The main principles for simulation of distributions (univariate at least) are the same as for the discreet distributions. Moreover, some densities may be conveniently and quite accurately approximated by means of discrete distributions. Further, we shall list the mentioned ideas without proofs, which can be found, in particular, in [25].

1. Inversion method. Let $F(x)$ be a continuous distribution function of a random value ξ, and $F^{-1}(y)$ be an inverse function

$$F^{-1}(y) = \inf\{x : F(x) = y,\ 0 \le y \le 1\}.$$

Then

a) if α_i $(i = 1, 2, \dots)$ are independent realizations of a uniformly distributed random value on $[0, 1]$, then $F^{-1}(\alpha_i) = \xi_i$ are independent realizations of a random value ξ;

b) if ξ_i $(i = 1, 2, \dots)$ are independent realizations of a random value ξ, then $F(\xi_i)$ are independent realizations of a random value, which is uniformly distributed on $[0, 1]$.

2. Inversion method for random vectors simulation. Let $\Xi = (\xi_1, \dots, \xi_s)$ be a random vector, $j_1, \dots, j_s$ be a rearrangement of numbers $1, \dots, s$, and conditional distribution functions of the components $\xi_1, \dots, \xi_s$ are known.

$F_1(x_{j_1}) = F_1(x_{j_1})$ is a distribution function of ξ_{j_1},

$F_2(x_{j_2}) = F(x_{j_2} \mid \xi_{j_1} = x_{j_1})$ is a distribution function of ξ_{j_2} provided that $\xi_{j_1} = x_{j_1}$,

$F_s(x_{j_s}) = F_s(x_{j_s} \mid \xi_{j_1} = x_{j_1}, \dots, \xi_{j_{s-1}} = x_{j_{s-1}})$ is a distribution function ξ_{j_s} under the conditions $\xi_{j_1} = x_{j_1}, \dots, \xi_{j_{s-1}} = x_{j_{s-1}})$.

Then if $\alpha_1^{(i)}, \dots, \alpha_s^{(i)}$, $i = 1, 2, \dots$, are independent realizations of a uniformly distributed on $[0, 1]$ random value, then $F_1^{-1}(\alpha_1^{(i)}), \dots, F_1^{-1}(\alpha_i^{(s)})$ are independent (for different i) realizations of a random vector $(\xi_{j_1}^{(i)}, \dots, \xi_{j_s}^{(i)})$ which coincides with an approximation of enumerated components with an initial vector Ξ .

3. Rejection method. Let $f(X)$, $X = (x_1, \dots, x_s)$ be a distribution density on $\mathbb{R}^s$. Then:

a) if (Ξ, η) is a random vector uniformly distributed on a set $a = \{(X, y) : X \in \mathbb{R}^s,\ 0 \le y \le cf(X)\}$, $c > 0$, then Ξ has a distribution density f;

b) if Ξ is a random vector with the density f and α is an independent random value with the uniform distribution on $[0, 1]$, then a random vector (Ξ, α) has a uniform distribution on a.

4. Composition method. Let Ξ have a density on $\mathbb{R}^s$ that can be represented in the form

$$f(X) = \int_{\mathcal{Y}} \varphi(dY)\, P(X,Y), \tag{2.7}$$

where φ is a probability measure on a set $\mathcal{Y}$, $\mathcal{Y} \in \mathbb{R}^r$, and $P(X,Y)$ is a density on $\mathbb{R}^s$ at every fixed Y. Then if vector $(\eta_1, \dots, \eta_r)$ is distributed in accordance with measure φ, and Ξ has distribution density $P(X/Y = (\eta_1, \dots, \eta_r))$, then $f(X)$ is also a distribution density.

It is evident that this statement is a consequence of the formulae from item 2. Although it may seem quite trivial, it can have many applications. Suffice it to say that any "combination" of distributions can be represented in the form of (2.7).

Besides these quite universal approaches to simulation that have been mentioned, there are many particular considerations which are based on density properties, both general and pertaining to separate specific distributions, and which are well known in probability theory. Among general properties one may consider the density transformation formula when substituting variables, the convolution formula, and particularly in the use of limit theorems and of majorants, etc. and others.

The problem of the optimal GDD algorithms for continuous density simulations is also considered in [62].

Example. [91]. Further, we shall consider the so-called pseudoinversion method as an example of an application and of some development of a standard simulation technique. Its main idea is the following: if $F_\xi(x)$ is a distribution function of a simulated random value ξ, and $F_\xi^{-1}(y)$ is inverse to it, then having a function $G(y)$ which is simply calculated and is close to $F^{-1}(y)$, one may improve the simulation algorithm ξ. For a strictly increasing G from the closeness of G to F^{-1}, it follows that $F(G(y))$ is close to y.

A random value β, such that $\xi G(\beta)$, should have a distribution close to the uniform distribution. Assume G increases monotonously. Then it is easy to see that the distribution function β, such that $\xi = G(\beta)$, is $H_\beta(y) = F(G(y))$. If the relevant derivatives $F'(x) = f(x)$ and $G'(x)$ exist, then the distribution density of β is $h(y) = (G(y))G'(y),\ 0 \le y \le 1$.

From the above, it follows that in the representation $h(y) = pu(y) + (1-p)r(y)$, where $u(y)$ is the density of the uniform distribution, p should be close to a unit (the closer G is to F^{-1}, the closer p is to unit), and $r(y) = (f(G(y))G'(y) - pu(y))(1-p)^{-1}$. Since $(1-p)$ is small, then the rare event, consisting in the simulation of $r(y)$, can be carried out by the selection method. To simulate the density

$$f(x) = \begin{cases} \sqrt{\frac{2}{\pi}}\, e^{-x^2/2}, & x \ge 0, \\ 0, & x < 0, \end{cases}$$

by this method, it is supposed in [24] that

$$G(y) = \left(-1,553 \ln(1-y^2)\right)^{1/2}, \quad 0 \le y \le 1,$$
$$h(y) = 0,985\, u(y) + 0,015\, r(y), \quad 0 \le y \le 1,$$

where $r(y) = \left[1,239\, y(1-y^2)^{-0,224}/(-1,553 \ln(1-y^2))^{1/2} - 0,985\right] 0,015$.

Here in 985 cases out of 1000, a uniform distribution is simulated, and only in 15 is a complex distribution with density $r(y)$ simulated. For constructing algorithms for the multiple simulation more effort is needed to increase the speed of a density $r(y)$ simulation.

We offer this example to show that rather a complicated reasoning is often required to increase the simulation speed. (A reader is supposed to be familiar with the simplest examples given, for example, in [36].)

3. Markov chain simulation

The Markov chain with discrete time and a finite or countable number of states is one of the simplest random processes. A definition and the study of the properties of that chain are available in [54].

First of all, we will describe a simulation technique for that chain which, in fact, can be taken as its definition. We shall try to give the minimum amount of information which will be necessary for further use.

Without loss of generality, we believe that time t takes integer values $t = 0, 1, 2, \ldots$ Consider a sequence of integer-valued random quantities $i(t)$, each of which takes one of the values $1, 2, \ldots, n$, where n is the number of states in which the chain can be found. Let it be said that $i(t)$ forms a Markov chain if $i(0)$ has a discrete distribution $p^0(1), \ldots, p^0(n)$, and, with $t > 0$, it has a discrete distribution $p(t,\, i(t-1) \to j)$, $j = 1, 2, \ldots, n$. A Markov chain is thus defined by a vector of the initial distribution $p^0 = (p^0(1), \ldots, p^0(n))$ and by a transition matrix $\mathcal{P}(t) = \|p(t,\, i \to j)\|_{i,j=1}^{n}$. If $\mathcal{P}(t)$ does not depend on t, a chain is called homogeneous.

A simulation algorithm for a random value $i(t)$ follows directly from the given definition. Given the found value $i(t-1)$, in order to receive $i(t)$, it is necessary to simulate a discrete distribution $p(t,\, i(t-1) \to 1)$, $p(t,\, i(t-1) \to 2)$,...,$p(t,\, i(t-1) \to n)$ (n may be infinite). Simulation methods have been discussed earlier.

As a result of simulation, we have a sequence (random) of numbers $i(0) \to i(1) \to \ldots$ whose behavior depends on a matrix structure $\mathcal{P}(t)$. A full analysis of this structure for the homogeneous case may be found in [40, vol.1].

At every instant of time t, a system is characterized by a vector of a state probability distribution $p(t) = (p_1(t), \ldots, p_n(t))$, where $p_j(t) = P\{i(t) = j\}$, $j = 1, \ldots, n$. As is known,

$$p(t+1) = p(t)\mathcal{P}(t), \quad \mathcal{P} = \|p(t,\, i \to j)\|$$

or

$$p(t+1) = p(0)\mathcal{P}^t = p(t)\mathcal{P}, \quad \mathcal{P} = \|p(i \to j)\| \tag{3.1}$$

in the homogeneous case.

Let us study this case in detail. Matrix $\mathcal{P}$ defines the probabilities of transition from a state with index i into a state with index j in one step (per time unit). From (3.1) it follows that the probabilities $p_{i,j}^{(m)}$, $m = 2, 3, \dots,$ $p_{i,j}^{(1)} = p(i \to j)$ of transition from a state with index i into a state with index j in m steps are the elements of a matrix

$$\left\| p_{i,j}^{(m)} \right\| = \mathcal{P}^m$$

and for all $m, n,$ $m = 1, 2, \dots;$ $n = 1, 2, \dots,$ the equality

$$p_{i,j}^{(m+n)} = \sum_k p_{i,k}^{(m)} p_{k,j}^{(n)}.$$

is valid.

A state with index k is said to be attainable from a state with index j if there exists $n \geq 1$ such that $p_{j,k}^{(n)} > 0$. For the reader's convenience, we would like to recall the following definition.

Definition. A state set C with indexes $j_1, \dots, j_k$ is called a closed set if no state outside C is attained from no state inside C.

If one of the states is closed, then it is called an absorbing state.

If a chain has no closed state, then it is called an irreducible chain.

Some necessary information concerning an absorbing state is given below.

Let $(\tilde{p}^0, \widehat{\mathcal{P}})$ be a Markov process with absorbing $n + 1$th state and

$$\tilde{p}^0 = \left(p_1^0, \dots, p_n^0, 0\right), \quad \mathcal{P} = \begin{pmatrix} \mathcal{P}, & Q \\ \tilde{0}, & 1 \end{pmatrix}, \quad Q = \begin{pmatrix} q_1 \\ \vdots \\ q_n \end{pmatrix} \quad \tilde{0} = (0, \dots, 0).$$

Then the equality (3.1) may be rewritten in the form:

$$\left(p(t+1), \, p_{n+1}(t+1)\right) = \left(p(t)\mathcal{P}, \, p(t)Q + p_{n+1}(t)\right), \tag{3.2}$$

where $p(t) = (p_1(t), \dots, p_n(t))$.

Another form of (3.2) is the equality

$$\left(p(t+1), \, p_{n+1}(t+1)\right) = \left(p(0)\mathcal{P}^{t+1}, \, p(0)\sum \mathcal{P}^j Q\right), \quad t = 0, 1, \dots$$

If a matrix norm $\mathcal{P}$ is less than 1, then $p_{n+1}(\infty) = p(0)q$. Here a vector q is the solution of the equation $q = \mathcal{P} + Q$. If $Q = 0$, then the largest eigenvalue $\mathcal{P}$ is equal to a unit, and either there exists a limit

$$\mathcal{P}(\infty) = \lim_{N \to \infty} \mathcal{P}^N, \tag{3.3}$$

and $\pi = p(0)\mathcal{P}(\infty)$ is the first eigenvector for a matrix $\mathcal{P}$, i.e. $\pi = \pi\mathcal{P}$, or for a certain k at each j, there exist the limits

$$\mathcal{P}_j = \lim_{N \to \infty} \mathcal{P}^{jN}, \quad \mathcal{P}_j \neq \mathcal{P}_k, \quad j \neq k, \quad i, j = 1, \dots, k-1.$$

In the latter case, the matrix $\mathcal{P}$ is called periodic. Thus, for irreducible and aperiodic $\mathcal{P}$, there exists a first eigenvector π which gives a distribution $\sum \pi_i = 1$, at least for the finite N.

From (3.3), it follows that π may be approximately calculated by calculating vector $p(o)$ iteration (the latter must not be orthogonal to π). The convergence rate of a vector $\pi_n = p(0)\mathcal{P}^n$ to π as n increases is defined by the second eigenvalue λ. An appropriate error is $O(\lambda^n)$.

Functionals of a stationary distribution π of the Markov chain can be estimated by its trajectory simulation and by calculating a certain estimate on it. Let a vector $h = (h_1, \dots, h_n)$ be given, and $i_0 \to i_1 \to \dots$ be the Markov chain trajectory . Then

$$\lim_{N\to\infty} \frac{1}{N} \sum_{k=0}^{N-1} h_{i_k} = (h, \pi). \tag{3.4}$$

If we consider the case of a finite or countable number of states, then each state, as a rule, is attainable in a finite number of steps from any other state . In this case, one state may be selected (let its number be k) and the chain trajectory can be divided into stages, i.e. parts which begin with index k but which do not contain this index in them. A chosen state becomes an absorbing state and here it turns out to be possible to apply the theoretical results obtained for estimates of the solution of linear algebraic equations systems, to the problem of estimating the eigenfunctions of the matrix $\mathcal{P}$.

Monte Carlo methods, which have been developed to solve such classical problems as the problems of linear algebra and mathematical physics, etc., are often found to be applicable to the problems of system's simulation, where they are of more practical importance than in solving traditional problems.

The next remark often turns out to be useful for spreading the known ways of simulation improvement to a wider class of problems.

Suppose that in a selected state space $\{1, 2, \dots, n\}$, the process is not a Markov process but, at any time t, the probability of transition from state i_t into state i_{t+1} depends on the indexes $i_{t-s}, i_{t-s+1}, \dots$ of states no more than at s previous to time t, i.e. the matrix $\mathcal{P}$ of the transition from a state with index i into a state with index j has the form

$$\mathcal{P} = \|p(i_1, \dots, i_s \to j)\| \,. \tag{3.5}$$

Here it is easy to show another state space, where the process will be a Markov process, and to construct the corresponding transition matrix. One possible way of introducing such a state space is the following. Let us consider the set $(i_1, \dots, i_s)$ to be new states. Altogether there are n^s state. If we suppose that

$$\tilde{p}(i_1, \dots, i_s \to j_1, \dots, j_s) = \begin{cases} p(i_1, i_2 \dots, i_s \to j_s), & \text{if } j_1 = i_2, \dots, j_{s-1} = i_s, \\ 0, & \text{if } (j_1 \neq i_2) \vee \dots \vee (j_{s-1} \neq i_s), \end{cases}$$

then the newly constructed random process with n^s states and transition matrix

$$\widehat{\mathcal{P}} = \|\tilde{p}(i_1, \dots, i_s \to j_1, \dots, j_s)\|^n_{i_k, j_k = 1}, \quad k = 1, \dots, s,$$

is the required Markov process.

It should be mentioned that many probabilistic models of real processes are described (often approximately) by using matrix (3.5).

The next, according to its complexity, process which is important for applications in simulation theory and practice, is the Poisson random process – the Markov process with continuous time which is the simplest model of a random flow of homogeneous events. Let t and T be two real numbers which will be interpreted as certain points of time $0 \le t < T$, and let $N(t,T) \ge 0$ be the number of events at the interval (t,T) ($N(t,T)$ is a whole number). Assume $r = T - t$. The following three properties define the Poisson process:

1. For nonintersecting intervals (t_1,t_2), $(t_3,t_4), \dots$, random values $N(t_1,t_2)$, $N(t_3,t_4), \dots$ are independent.

2. $N(t,T)$ is distributed as $N(0,T-t)$, i.e. the number of events depends only on the interval length (t,T) and not on the initial point.

3. The probability, that more than one event will occur in an arbitrary small period of time r, is $O(r)$ and

$$P\{N(0,r) = 1\} = \lambda(t)r + O(r),$$

where $\lambda(t) \ge 0$ and $\int_0^T \lambda(t)\,dt \to \infty$ for $t \to \infty$.

From these properties, it is easy to obtain (see, e.g., [105]) that the period of time r between the occurrences of two successive events is distributed exponentially with parameter $\lambda(t)$, i.e. has the distribution density

$$\varphi(x) = \lambda(x)\, e^{\int_0^x \lambda(t)\,dt}, \quad x \ge 0, \tag{3.6}$$

and the events flow described by the Poisson process may be simulated by successive simulation of random periods of time whose lengths have density (3.6).

Another (approximated) way of simulation is based on Property 3: at a sufficiently small period of time (t,T) with length r, the event occurs with probability $\lambda(t)r$.

As for the exponential distribution, a lot of ways exist for its simulation. At a constant λ, the inversion formula yields

$$\xi = -(1/\lambda)\ln\alpha,$$

where α is uniformly distributed on $[0,1]$. Multiple calculations of a logarithm are undesirable for a substantial calculation. That's why other special methods are used [24,25]. If $\lambda(t)$ depends on time in a complicated way, then it is not sensible to use the inversion formula. For that case, a series of methods may be suggested, such as the selection method, which make the procedure considerably less labour-consuming.

CHAPTER 2. DISCRETE MARKOV CHAINS

One of the most important techniques in the simulation of complex stochastic systems is to construct an embedded Markov chain. As using of a computer in this case is inevitable, we deal with a finite number of states. Although the investigation of general state space is often required for the successful discretization of the problem, an algorithm is written in terms of discrete or finite Markov chains. Besides, the case of a discrete state space enables us to introduce and to illustrate the basic equations, standard estimators and ways of choice of experimental designs. Though a generalization for the case of an arbitrary (with some restrictions, typical for the theoretical probabilistic approach) state space is not always trivial, it is still performed with a unified scheme and will be discussed in chapter 4. The excellent monographs [54,55], dealing with theoretical probabilistic problems, are devoted to discrete Markov chains. However, the problem of constructing estimators is only slightly touched on in these books, and the methods of experimental design are even not mentioned. On the other hand, many of the problems, which are thoroughly investigated in this chapter, are touched upon in papers on Monte Carlo methods. But the problems of estimating finite horizon performance, which are rather typical for simulation, haven't been practically considered in monographs, and the methods suggested in journal articles, have still not been systematized. In this chapter, the methods of estimation and experimental design, suggested in [22,29,51,74], are considered from a uniform point of view.

We will restrict our considerations to two classes of finite Markov chains: absorbing (and corresponding to them, breaking off) chains and positively recurrent aperiodic ones. An investigation of arbitrary chains can be reduced to studying these two chain classes (see [54]). For convenience, some necessary definitions are given in Section 1.

Note that the basic relation representing the distribution of probability states on the next step by the distribution of probability states on the previous step, leads to the corresponding recurrent relation for estimators. For breaking off Markov chains, this relation and its dual result, in the limit, in a pair of dual equations and the corresponding representations for variances and covariances. This enables us to obtain a suitable representation for variances and covariances of estimators from finite segments of Markov chains. It is important that such a representation is expressed in terms of functionals of the path of the chain.

For the simulation of ergodic Markov chains, the important point is the availability of moments of regeneration, i.e. moments with segments of the chain between them which are probabilistic replicas of each other. For finite Markov chains, such moments are the jumping moments into any fixed state. Using this, we will be able to get a representation for variances of functionals based on the investigation of embedded breaking off chains.

Another way of ergodic chain simulation is to partition them into segments of a fixed length. It enables us to simulate not only the main chain but also embedded chains of a fixed length with the same primary state. Representations of variances and covariances, mentioned above, can be used for these chains, too.

One way to increase efficiency of simulation is to use unbiased estimators in the form of linear functionals of auxiliary Markov chains called fictitious chains.

Representations for variances and covariances in the case of breaking off chains and the unique positive linear functional, permit us to construct the estimators with zero variance, which, however, are based on the exact value of the functional at any initial distribution and cannot be directly applied. Nevertheless, these representations allow the using of a prior information to increase the simulation efficiency and previous experimental simulation results in the sense of the sequential approach. The corresponding method of simulation is called the method of importance sampling.

Instead of fictitious chain's simulation, one may use a branching random process controlling the number of parallel simulated independent Markov chains on the basis of splitting and Russian roulette approaches. The next chapter is devoted to this problem.

1. Notations and Statement of the Problem

In this section and in the following one, we will consider discrete Markov chains. For the convenience of presentation, we will introduce a number of standard definitions and denotations to be used. Let $\mathcal{X} = \{1, 2, \ldots, n\}$ or $\{1, 2, \ldots\}$, $\mathcal{X}' = \{0\} \cup \mathcal{X}$.

Definition 1.1. A sequence of discrete random quantities $\xi_0, \xi_1, \ldots$ with values in the set $\mathcal{X}(\mathcal{X}')$, such that

$$P\{\xi_s = x_s \mid \xi_{s-1} = x_{s-1}, \ldots, \xi_0 = x_0\} = P\{\xi_s = x_s \mid \xi_{s-1} = x_{s-1}\}$$

is called a discrete Markov chain (with the state space $\mathcal{X}(\mathcal{X}')$).

A matrix

$$P_{(s)} = \Big(P\{\xi_s = x \mid \xi_{s-1} = y\} \Big)$$

is called a matrix of the transition probabilities at the sth step, and for $P_{(s)} \equiv P$, a matrix P is simply called a transition matrix.

In the case of a space $\mathcal{X}$, we will use the notation Q and $Q_{(s)}$ for the matrix of transition probabilities and for the matrix of transition probabilities at the sth step, respectively.

If $P_{(s)} \equiv P$ (or in case of a space $\mathcal{X}$, $Q_{(s)} \equiv Q$), then a Markov chain is called a homogeneous Markov chain.

A sequence of values of random variables $\xi_0, \xi_1, \ldots$, which will be denoted in the same way as random variables themselves, is called the path of a Markov chain. Further, we will consider only positively recurrent aperiodic chains (such chains are often called regular chains) and absorbing chains.

Determine the indicated chain types by introducing a series of standard auxiliary notions.

When studying paths of infinite length, one is usually constrained by homogeneous Markov chains. Further, it will be clear from the context when homogeneous chains are meant. So the word "homogeneous" will be omitted. For homogeneous Markov chains, a transition probability for a k steps matrix is equal to P^k and Q^k, respectively.

It is well known (see [54]) that the investigation of homogeneous discrete Markov chains can be reduced to the investigation of ergodic chains with a single class of equivalence.

A state $\{x\}$ of a homogeneous Markov chain is called a reciprocal state if $\sum_{k=1}^{\infty} P^k(x,x) = 1$. A state (x) of a Markov chain is called an absorbing one if $P(x,x) = 1$.

Definition 1.2. A homogeneous Markov chain is called an absorbing chain if, for any initial state, a path jumps into any absorbing state with the probability equal to 1.

Further, we will consider only absorbing Markov chains with a unique absorbing state $\{0\}$.

We will consider states x and y to be communicating if

$$\sum_{k=1}^{\infty} P^k(x,y) > 0, \quad \sum_{k=1}^{\infty} P^k(y,x) > 0.$$

Consider chains with a unique class of equivalence, i.e. chains with all states communicating.

Definition 1.3. A chain is called regular if all its states are communicating and

$$P^k(x.y) \to \pi(y)$$

for $k \to \infty$ for any $x, y \in \mathcal{X}(\mathcal{X}')$, where $\sum \pi(y) = 1$, $\pi(y) \geq 0$, is a vector called a stationary distribution.

It is known that a chain with a finite set of states is regular if and only if there exists such a number N that

$$P^N(x,y) > 0,$$

for any $x, y \in \mathcal{X}(\mathcal{X}')$.

We will be interested to know how often a Markov chain hits into this or that state. Generally speaking, as the value of various states jumping is different from the viewpoint of system development (represented here by the Markov chain), weights $h(x)$, where $h(x)$, $x \in \mathcal{X}'$ is a given real-value function, $h: \mathcal{X}' \to R^1$ are assigned to these states.

Definition 1.4. A quantity

$$J_{(N)} = \sum_{k=0}^{N} \sum_{y \in \Omega} P^{(k)}(y) h(y),$$

where $\{P^{(k)}(y)\}$ is the distribution of a random value ξ_k, N is a given number, $\Omega = \mathcal{X}$ or $\mathcal{X}'$, is called a linear functional (of distributions) of the Markov chain (in the transition regime).

A considerably more general definition of a quantity $J_{(N)}$, which will be also suitable for nonhomogeneous Markov chains, can be introduced

$$J_{(N)} = \sum_{k=0}^{N} P^T R(1) \dots R(k)\, h(k),$$

where $(R(k))(x,y) = P_{(k)}(x,y)\,a_{(k)}(x,y)$, $x,y \in \Omega$, $\Omega = \mathcal{X}$ or $\mathcal{X}'$, $a_{(k)}(x,y)$ are given real coefficients, $h^T(k) = (h(x;k))_{\mathcal{X}\in\Omega}$ and $h(x;k)$ are given real functions.

Quantities $h(x;k)$, $a_k(x,y)$ are determined by the substantial interpretation of the Markov chain.

In the case of prolonging the functioning of a chain, the problem of estimating stationary distribution functionals arises. If such a distribution exists, one of the advantages of its application is it's independence from the length of a path (N) and the initial distribution.

Definition 1.5. A quantity

$$J_{st} = \sum_{y\in\Omega} \pi(y)\,h(y),$$

$\Omega = \mathcal{X}$ or $\mathcal{X}'$, is called a linear functional of the stationary distribution of a Markov chain.

Definition 1.6. A quantity

$$J = \lim_{N\to\infty} J_{(N)} = \lim_{N\to\infty} \sum_{k=0}^{N} \sum_{y\in\Omega} P^{(k)}(y)\,h(y),$$

is called a linear functional of an absorbing Markov chain with the unique absorption state $\{0\}$.

Since the considered chain with probability equal to 1, hits into state $\{0\}$, then the definition is correct if $\sum |h(y)| < \infty$.

The problem of this chapter and of the next one, is to construct and investigate linear functionals $J_{(N)}$, J_{st}, and J statistical estimators along the paths of a Markov chain.

2. Estimation of the functional $J_{(N)}$

Let $\xi_0, \xi_1, \ldots, \xi_N$ be an inhomogeneous, generally speaking, Markov chain with the initial distribution p and transition matrices $P_{(l)}$, $l = 1, 2, \ldots, N$.

Consider the estimate problem of linear functional $J_{(N)}$ defined in (1.1). In the case of $a_{(k)}(x,y) \equiv 1$, the following estimator can be introduced:

$$\widehat{J}_{(N)} = \sum_{k=0}^{N} h(\xi_k; k). \tag{2.1}$$

This estimator has a simple probabilistical sense: when a chain jumps into a state x at the kth step, a quantity $h(x;k)$ is added to the value of an estimator. It is clear that the estimator $\widehat{J}_{1(N)}$ is unbiased.

Instead of the chain $(p, P_{(l)})_{l\in 1:N}$, some alternative chain, called a fictitious chain, can be simulated. Consider the following estimator

$$\widehat{J}_{1(N)} = \sum_{k=0}^{N} \Big(p(\tilde{\xi}_0)\,/\,\tilde{p}(\tilde{\xi}_0)\Big) \prod_{l=1}^{k} v_l(\tilde{\xi}_{l-1}, \tilde{\xi}_l)\,h(\tilde{\xi}_k; k), \tag{2.2}$$

where $\tilde{\xi}_0, \dots, \tilde{\xi}_N$ is the path of the chain $(\tilde{p}, \tilde{P}_{(l)})_{l \in 1:N}$,

$$v_l(\tilde{\xi}_{l-1}, \tilde{\xi}_l) = R_{(l)}(\tilde{\xi}_{l-1}, \tilde{\xi}_l) \Big/ \left(\tilde{P}_{(l)}(\tilde{\xi}_{l-1}, \tilde{\xi}_l)\right).$$

For $\tilde{p} = p$, $\tilde{P}_{(l)} = P_{(l)}$, $a_{(l)}(x, y) \equiv 1$, $x, y \in \Omega$, $l = 1, 2, \dots, N$ estimators (2.2) and (2.1) coincide.

Such estimators were introduced by von Neumann and Ulam (see [29]) and are widely used in papers on the Monte Carlo method with the supposition of chain homogeneity and at $N \to \infty$.

Denote by pq a vector with elements which are the products of vector elements p and q. If $p = q$, then pq is written as p^2.

Denote by P_{sq}, P_{sq}/Q a matrix with elements $P^2(x, y)$ and $P^2(x, y)/Q(x, y)$, respectively.

Let a fictitious chain possess the following property: from $P_{(l)}(x, y) > 0$ it follows that $\tilde{P}_{(l)}(x, y) > 0$

$$l = 1, 2, \dots, N\,; \quad x, y \in \Omega, \tag{2.3}$$

from $p(x) > 0$, it follows that $\tilde{p}(x) > 0$, $x \in \Omega$.

If, in the product

$$\prod_{j=1}^{k} v_j(\xi_{j-1}, \xi_j)\; P(\xi_0)/\tilde{P}(\xi_0) = \prod_{j-1}^{k} \frac{R_{(i)}(\xi_{j-1}, \xi_j)}{\tilde{P}_{(j)}(\xi_{j-1}, \xi_j)},$$

a multiplier in the numerator is equal to zero, then the product is assumed equal to zero. But if condition (2.3) is fulfilled, the probability of this event equals zero.

Note that

$$J_{(N)} = (p, \varphi(N)), \tag{2.4}$$

where $\varphi(N)$ is defined by relations:

$$\begin{aligned} &\varphi(0) = h(N), \\ &\varphi(k+1) = R_{(N-k)}\,\varphi(k) + h(N-k-1), \quad N \geq k+1 \geq 1. \end{aligned} \tag{2.5}$$

Indeed, consecutively finding $\varphi(1), \dots, \varphi(N)$, we get

$$\varphi(N) = h(0) + R_{(1)}h(1) + \ldots + R_{(1)} \ldots R_{(N)}h(N).$$

Multiplying both sides of this equality by p^T from the left, we obtain (2.4). Define quantities

$$\begin{aligned} &\bar{\varphi}(x; 0) = \varphi(x; 0) = h(x; N), \quad x \in \Omega, \\ &\bar{\varphi}(x; k+1) = \sum_{y \in \Omega} \zeta_{(N-k)}(x, y)\; v_{(N-k)}(x, y)\; \bar{\varphi}(y; k) + h(x; N-k-1), \\ &N \geq k+1 \geq 1, \end{aligned} \tag{2.6}$$

where quantities $\bar{\varphi}(x;k)$ and $\zeta_{(N-k)}(x,y)$ are determined at $\tilde{\xi}_{(N-k)} = x$ and $\zeta_{(N-k)}(x,y) = 1$ if $\tilde{\xi}_{(N-k)} = y$ and 0 otherwise, $\tilde{\xi}_0, \ldots, \tilde{\xi}_N$ is the path of the chain $(\tilde{p}, \tilde{P}_{(l)})_{l \in 1:N}$.

The following theorem gives us nonbias of estimator $\widehat{J}_{1(N)}$ defined by formula (2.2) and an occurrence for its variance.

Theorem 2.1. *Let $(p, P_{(l)})_{l\in 1:N}$ be a discrete Markov chain. If condition (2.3) is fulfilled, then the estimator $\widehat{J}_{1(N)}$ given by formula (2.2) is an unbiased estimator of the functional $J_{(N)}$ and*

$$E\widehat{J}^2_{1(N)} = (p^2/\tilde{p},\ E(\bar{\varphi}(N))^2) = \sum_{k=0}^{N} \left(l(k),\ 2\varphi(N-k)\,h(k) - h^2(k)\right),$$

where

$$\varphi(k) = E\bar{\varphi}(k), \tag{2.7}$$

quantities $\varphi(x;k)$, $k = 0,1,\ldots,N$ are defined by (2.5) and

$$\begin{aligned} &E\bar{\varphi}^2(0) = h^2(N),\\ &E(\bar{\varphi}(k+1))^2 = \left(R_{sq}(N-k)/\tilde{P}(N-k)\right)\, E(\bar{\varphi}(k))^2 + \\ &\qquad\qquad + 2\varphi(k+1)\,h(N-k-1) - h^2(N-k-1),\\ &l(0) = p^2/\tilde{p}, \quad l^T(k+1) = l^T(k)\left(R_{sq}(k+1)/\tilde{P}(k+1)\right),\\ &N-1 \geq k \geq 0. \end{aligned} \tag{2.8}$$

Before proceeding to the proof of the theorem, we note that if all the computations necessary for finding the variances of the estimator $\widehat{J}_{1(N)}$ can be fulfilled, then there is no need for chain simulation because the computation of $J_{(N)}$ by formulas (2.4) and (2.5) is simpler than variance computation. But this result is useful for the theoretical analysis of the required volume of computations and for constructing optimal fictitious chains.

Proof. Quantities $\zeta_{(N-k)}(x,y)$ reflect the transitions of the chain to the step with number $N-k$ and, directly from their definition, it follows that

$$\begin{aligned} &E\zeta_{(N-k)}(x,y) = P\left\{\tilde{\xi}_{N-k} = y \mid \tilde{\xi}_{N-k-1} = x\right\} = \tilde{P}_{(N-k)}(x,y),\\ &E\zeta_{(N-k)}(x,y)\,\zeta_{(N-k)}(x,z) = 0 \quad (y \neq z),\\ &E\zeta^2_{(N-k)}(x,y) = \tilde{P}_{(N-k)}(x,y). \end{aligned}$$

Calculating the expectation of the left and right sides of (2.6), we get (2.7). Note that

$$\widehat{J}_{1(N)} = \sum_{x\in\Omega} \zeta(x)\,\bar{\varphi}(x;N), \tag{2.9}$$

where $\zeta(\tilde{\xi}_0) = p/\tilde{p}$ and $\zeta(x) = 0, \quad x \neq \xi_0$.

Taking the expectation from both sides of (2.9) and taking into account the relations $E\bar{\varphi}(N) = \varphi(N)$ and $E\zeta = p$, we have

$$E\widehat{J}_{1(N)} = J_{(N)}.$$

Taking the square of both sides of (2.9) and calculating their mathematical expectations, we have

$$E\widehat{J}^2_{1(N)} = \sum_{x\in\Omega} \left(p^2(x)/\tilde{p}(x)\right) E\left(\bar{\varphi}(x;N)\right)^2 = \left(p^2/\tilde{p},\ E\bar{\varphi}^2(N)\right). \tag{2.10}$$

Calculating the mathematical expectations from both sides of the squares of (2.6), we have (2.8). Substituting (2.8) in (2.10), we obtain

$$E\widehat{J}^2_{(N)} = \sum_{k=0}^{N} \left(l(k),\ 2\varphi(N-k)h(k) - h^2(k)\right).$$

The proof is complete.

The formulas for the variance obtained in Theorem 2.1 enable us to construct estimators with zero variance in the case of $h(x;k) \geq 0$.

Consider a class of fictitious chains of the following type:

$$\begin{aligned} &\tilde{p} = p\, q(0)/(p, q(0)), \\ &\tilde{p}_{(l)}(x,y) = R_{(l)}(x,y)\, q(y;l) \Big/ \sum_{z\in\Omega} R_{(l)}(x,z)\, q(z;l)\,, \quad l = 1,\dots,N, \end{aligned} \tag{2.11}$$

where $q(l) = (q(x;l))_{x\in\Omega}$, $l = 0,1,\dots,N$ are arbitrary nonnegative vectors.

Theorem 2.2. *In a class of fictitious chains of the above type satisfying condition (2.3), the minimum of $\mathcal{D}\widehat{J}_{1(N)}$ is attained if and only if*

$$q(l) = C_l|\varphi(N-1)|, \tag{2.12}$$

where C_l are arbitrary positive constants, $l = 0,1,\dots,N$.

With the above choice of $q(l)$, if $h(x,k) \geq 0$ $(x \in \Omega,\ k = 0,1,\dots,N)$, then

$$\begin{aligned} &\tilde{p}(x) = p(x)\,\varphi(x;N)/J_{(N)}, \\ &\tilde{P}_{(l)}(x,y) = R_{(l)}(x,y)\,\varphi(y;M-1)\,/\left[\varphi(x;N-l+1) - h(x;l-1)\right], \end{aligned} \tag{2.13}$$

$l = 1,2,\dots,N$, $\mathcal{D}\widehat{J}_{1(N)} = 0$ and $\widehat{J}_{1(N)} \equiv J_{(N)}$ on any chain path.

Proof. Begin with the last statement. Let $h(x;k) \geq 0$. Substituting expressions (2.11) and (2.12) in $\widehat{J}_{1(N)}$, we have

$$\widehat{J}_{1(N)} = \sum_{k=0}^{N} \frac{J_{(N)}}{\varphi(\xi_0;N)} \prod_{l=1}^{k} \frac{\varphi(\tilde{\xi}_{l-1};N-l+1) - h(\tilde{\xi}_{l-1};l-1)}{\varphi(\tilde{\xi}_l;N-1)}\ (\tilde{\xi}_k;k).$$

In this expression, it is assumed that, for $k = 0$, a multiplier at $J_{(N)}/\varphi(\tilde{\xi}_0; N)$ is equal to $h(\tilde{\xi}_0; 0)$. Replace the multiplier $h(\tilde{\xi}_k; k)$ by $\varphi(\tilde{\xi}_k; N-k) + (h(\tilde{\xi}_k; k) - \varphi(\tilde{\xi}_k; N-k))$. Then the right side takes the form:

$$\frac{J_{(N)}}{\varphi(\tilde{\xi}_0; N)} \left[\sum_{k=1}^{N} (v_{k-1} - v_k) + h(\tilde{\xi}_0; 0) \right],$$

where

$$v_k = \prod_{l=1}^{k} \frac{\varphi(\tilde{\xi}_{l-1}; N-l+1) - h(\tilde{\xi}_{l-1}; l-1)}{\varphi(\tilde{\xi}_l; N-1)} \left(\varphi(\tilde{\xi}_k; N-k) - h(\tilde{\xi}_k; k) \right),$$
$$k \geq 1,$$
$$v_0 = \varphi(\tilde{\xi}_0; N) - h(\tilde{\xi}_0; 0).$$

Note that in the expression for v_N, the last multiplier equals to

$$\varphi(\tilde{\xi}_N; 0) - h(\tilde{\xi}_N; N) = 0$$

so $v_N = 0$. It follows that

$$\widehat{J}_{1(N)} = \frac{J_{(N)}}{\varphi(\tilde{\xi}_0; N)} \left(v_0 + h(\tilde{\xi}_0; 0) \right) = J_{(N)},$$

i.e. the estimator $\widehat{J}_{1(N)}$ does not depend on the path of a chain, and its variance is equal to zero.

Now let $\widehat{J}_{1(N)}(q)$ be an arbitrary estimator of the type (2.2) with a fictitious chain of the type (2.11) satisfying condition (2.3). The nonbias of the estimator $\widehat{J}_{1(N)}(q)$ has been proved in Theorem 2.1. Moreover, according to this theorem,

$$E\widehat{J}^2_{1(N)}(q) = \left(p^2/\tilde{p},\ E(\bar{\varphi}(N))^2\right),$$

where

$$\begin{aligned} &E\left(\varphi(0)\right)^2 = \left(\varphi(0)\right)^2 = h(N), \\ &E\left(\bar{\varphi}(x; k+1)\right)^2 = \sum_y \frac{R(x, y; N-k)}{q(y; N-k)} \sum_z R(x, z; N-k)\, q(z; N-k)\ E\left(\bar{\varphi}(y; k)\right)^2 + \\ &\qquad + 2\varphi(x; k+1)\, h(x; N-k-1) - h^2(x; N-k-1), \\ &N \geq k+1 \geq 1, \quad R(x, z; N-k) = R_{(N-k)}(x, z). \end{aligned} \tag{2.14}$$

Let the vectors $q(N), \dots, q(N-k+1)$ be fixed. Then the quantities $E\left(\bar{\varphi}(x,0)\right)^2, \dots,$ $E\left(\bar{\varphi}(x,k)\right)^2$ for any $x \in \Omega$ are also fixed. Let us consider the problem of which values of $q(N-k)$ quantity $E\left(\bar{\varphi}(x,k+1)\right)$ is minimal with a fixed x. Consider the Cauchy-Schwarz inequality in the form

$$\sum_{y\in\Omega} d(y)\, b(y)/\beta(y) \sum_{y\in\Omega} b(y)\, \beta(y) \geq \left(\sum_{y\in\Omega} b(y)\sqrt{d(y)}\right)^2,$$

the equality is attained if and only if $\beta(y) = \text{const}\sqrt{d(y)}$ at any fixed nonnegative $b(y)$, $d(y),\ y \in \Omega$.

Set

$$d(y) = E\left(\bar{\varphi}(y;k)\right)^2, \quad b(y) = R(x,y;N-k),$$
$$\beta(y) = \dot{q}(y;N-k).$$

Then relation (2.14) assumes the state

$$E\left(\bar{\varphi}(x;k+1)\right)^2 = \sum_{y} d(y)\, b(y)/\beta(y) \sum_{y} b(y)\,\beta(y) + c,$$

where c does not depend on $\beta = q(N-k)$. So the minimum of $E\left(\bar{\varphi}(x;k+1)\right)$ at any fixed $x \in \mathcal{X}$ and fixed $q(N), \dots, q(N-k+1)$ is attained if and only if

$$q(y;N-k) = \left[E\left(\bar{\varphi}(x,k+1)\right)^2\right]^{1/2}\text{const}.$$

Applying the performed reasoning at $k = 1, 2, \dots, N$, we have

$$\min\, E\left(\bar{\varphi}^{(N)}\right)^2 = \left(\varphi^{(N)}\right)^2$$

which is attained if and only if

$$q(N) = C_N|\varphi(0)|, \dots, q(1) = C_1|\varphi(N-1)|, \tag{2.15}$$

where $C_1, \dots, C_N$ are arbitrary positive constants.

It remains to define $q(0)$. If (2.15) is valid

$$E\widehat{J}^2_{1(N)}(q) = \left((p, q(0))p/q(0),\ \varphi^2(N)\right),$$

then applying the Cauchy-Schwarz inequality in the way shown above, we have that the minimum of $q(0)$ is attained if and only if $q(0) = C_0|\varphi(N)|$, where C_0 is an arbitrary positive constant.

Theorem 2.2 can be used to improve the statistical simulation efficiency in the following way. First, an approximate value of $\varphi(N-k)$ derived, for example, from analytical calculations for a chain with amplified states can be used as $q(k)$. Second, a consecutive approach can be used: running a series of experiments, we get the estimators of $\varphi(N-k)$ (we will denote them by $\hat{\varphi}(N-k)$) and set $q(k) = \hat{\varphi}(N-k)$, $k = 0, \dots, N-1$, in a new series of experiments. The results of this new series can be used to improve the next series efficiency, etc.

Let a number of experiments be performed (i.e. independent paths of the chain (p, P) be built up) and confidence intervals

$$\max_{x\in\Omega} \min \left\{ \frac{\varphi(x;k)}{\hat{\varphi}(x;k)}, \frac{\hat{\varphi}(x;k)}{\varphi(x;k)} \right\} \le 1 + \varepsilon_{N-k} \tag{2.16}$$

with probability γ obtained in the standard way.

Theorem 2.3. *If $h(x;k) \ge 0$, $x \in \Omega$, $k = 1, \dots, N$, then if conditions (2.3) and (2.16) are valid, the variance of the estimator $\widehat{J}_{(N)}$ of the chain (2.12) with $q(k) = \hat{\varphi}(N-k)$ satisfies the inequality*

$$\mathcal{D}\widehat{J}_{1(N)} \le \left[\prod_{k=0}^{N} (1+\varepsilon_k)^2 - 1\right] \widehat{J}^2_{(N)}. \tag{2.17}$$

Proof. By the definition of expectation and inequality (2.16), we have

$$
\begin{aligned}
&E\widehat{J}^2_{1(N)} = \\
&= \sum_{k=0}^{N} \sum_{x_0,\dots,x_k\in\Omega} \frac{p^2(x_0)\, R_1^2(x_0,x_1)\dots R_k^2(x_{k-1},x_k)}{\tilde{p}(x_0)\, \tilde{P}_1(x_0,x_1)\dots \tilde{P}_k(x_{k-1},x_k)}\, h^2(x_k;k) = \\
&= \sum_{k=0}^{N} \sum_{x_0,\dots,x_k\in\Omega} \frac{p(x_0)}{\hat{\varphi}(x_0,N)}\, R_1(x_0,x_1)\dots R_k(x_{k-1},x_k)\times \\
&\times \sum_{y\in\Omega} \frac{R_1(x_0,y)\,\hat{\varphi}(y;N-1)}{\hat{\varphi}(x_1;N-1)} \dots \sum_{y\in\Omega} \frac{R_{k-1}(x_{k-1},y)\,\hat{\varphi}(y;N-k)}{\hat{\varphi}(x_k;N-k)}\, h^2(x_k,k) \le \\
&\le \sum_{k=0}^{N} \sum_{x_0,\dots,x_k\in\Omega} \prod_{l=0}^{N} (1+\varepsilon_l)^2 \times \\
&\times \frac{p(x_0)}{\varphi(x_0,N)} R_1(x_0,x_1)\dots R_k(x_{k-1},x_k) \sum_{y} \frac{R_1(x_0,y)\,\varphi(y;N-1)}{\varphi(x_1;N-1)} \times \\
&\times \sum_{y} \frac{R_k(x_{k-1},y)\,\varphi(y;N-k)}{\varphi(x_k;N-k)} h^2(x_k,k) = \\
&= \prod_{l=0}^{N} (1+\varepsilon_l)\, E\left(\widehat{J}^*_{1(N)}\right)^2,
\end{aligned}
$$

where an estimator from Theorem 2.2 with the variance equal to zero, is denoted by $\widehat{J}^*_{1(N)}$. As $\widehat{J}^{*2}_{1(N)} = J^2_{(N)}$, then the estimator (2.17) and, hence, the theorem is proved.

Consider the problem of the simultaneous estimation of several functionals along the same chain path. The quantity

$$\sum_{v=1}^{m} a_v \mathcal{D}\, \widehat{J}_{1(N)}(h^{(v)}) \tag{2.18}$$

can be an estimator criterion where $\widehat{J}_{1(N)}(h^{(v)})$ denotes the estimator (2.3) for $h = h^{(v)}$ and a_v are the given nonnegative coefficients.

In this case, there are no estimators with zero variance, and the quantities $q(k)$ producing a minimum of the value of (2.18) are given in the following theorem.

Theorem 2.4. *Under the conditions of Theorem 2.2, the minimum of quantity (2.18) is attained if and only if*

$$q(x,k) = C_k \sum_{v=1}^{m} \left\{ E\left(\bar{\varphi}(x; N-k; h^{(v)})\right)^2 a_v \right\}^{1/2},$$

where C_k are arbitrary positive constants and $\bar{\varphi}(x; N-k; h^{(v)})$ is a quantity of $\bar{\varphi}(x; N-k)$ estimated at $h = h^{(v)}$.

The proof of this theorem and that of Theorem 2.2 are similar. Note that instead of (2.7), we will use the relation

$$\begin{aligned}
&\sum_{v=1}^{m} a_v\, E\left(\bar{\varphi}(x; k+1; h^{(v)})\right)^2 = \\
&= \sum_{v=1}^{m} a_v \left\{ \sum_y \frac{R(x,y;N-k)}{q(y;N-k)} \sum_z R(x,z;N-k)\, q(z;N-k) \right\} \times \\
&\times \left\{ E\left(\bar{\varphi}(x;k;h^{(v)})\right)^2 + 2\varphi(x;k+1;h^{(v)})\, h^{(v)}(x;N-k-1) - \right. \\
&\left. - \left(h^{(v)}(x;N-k-1)\right)^2 \right\},
\end{aligned}$$

which is the result of summing up the equalities (2.8) corresponding to $h = h^{(1)}, \ldots, h^{(m)}$.

If the requirements for the estimators precision are high, then the estimation of several functionals along the same chain path will be less efficient (there is no way of obtaining estimators with zero variance) than estimation of each functional along the path of a specially chosen chain.

Now we will consider the estimators of the functionals of absorbing Markov chains. Let an investigated chain and a functional be homogeneous: $P_{(l)} \equiv P$, $h(l) \equiv h$, $l =$

$1, 2, \dots, N$. Also let $a_l(x,y) \equiv a(x,y) \geq 0$, $l = 1, 2, \dots, N$. For such chains the results obtained above for the estimator of $J_{1(N)}$ are valid. A large number of unbiased estimators based on the absorption property is known for absorbing chains. It is known that, in the general case, there is not one of these estimators that can be regarded as the best one [29]. Therefore we will consider only the simplest one

$$\widehat{J}_{2(N)} = \frac{p(\tilde{\xi}_0)}{\tilde{P}(\tilde{\xi}_0)} \frac{R(\tilde{\xi}_0, \tilde{\xi}_1)}{\tilde{P}_1(\tilde{\xi}_0, \tilde{\xi}_1)} \cdots \frac{R(\tilde{\xi}_{k-1}, \tilde{\xi}_k)}{\tilde{P}_k(\tilde{\xi}_{k-1}, \tilde{\xi}_k)} \frac{h(\tilde{\xi}_k)}{\tilde{g}(\tilde{\xi}_k; k)}, \tag{2.19}$$

where $\tilde{g}(x_k; k) = 1 - \sum_{y \in \mathcal{X}'} \tilde{P}_k(x, y)$, k is an index of the last step before absorption. If absorption has not occurred for N steps, then set $\widehat{J}_{2(N)} = 0$.

As a transition from $\tilde{\xi}_k$ to the absorption state takes place, then with the probability equal to 1, $\tilde{g}(\tilde{\xi}_k; k) = 1 - \sum_y \tilde{P}(\tilde{\xi}_k, y) > 0$. Suppose that

$$\tilde{g}(x; k) > 0, \quad \text{at} \quad k = 1, 2, \dots, N \quad \text{if} \quad g(x) > 0. \tag{2.20}$$

We will assume if $a = 0$, $b = 0$, then $a/b = 0$. Let

$$\begin{aligned} &\bar{\varphi}(x; 0) = h(x)\, \nu(x) / \tilde{g}(x); \\ &\bar{\varphi}(x; k+1) = \frac{1 - \nu(x)}{1 - \tilde{g}_k(x)} \sum_y \zeta_k(x, y)\, v_k(x, y)\, \bar{\varphi}(y; k) + h(x)\, \nu(x) / \tilde{g}_k(x), \end{aligned} \tag{2.21}$$

where $\nu(x) = 1$ with probability $g(x)$, and 0 with probability $1 - g(x)$, $\zeta_k(x, y) = 1$ with probability $\tilde{P}_k(x, y)$, and 0 with probability $1 - \tilde{P}_k(x, y)$, $v_k(x, y) = R(x, y) / \tilde{P}_k(x, y)$, $\tilde{g}_k(x) = \tilde{g}(x, k)$.

Note that $\bar{\varphi}(x, y) = \widehat{J}_{2(k)}$ for $\tilde{p}(y) = \delta_{xy}$, $y \in \mathcal{X}'$ and recurrent formulas (2.21) reflect the change of the functional $\widehat{J}_{2(k)}$ at the transition from k to $k+1$.

The calculation of expectations of both sides of (2.21) and their squares leads to the following analogue of Theorem 2.1.

Theorem 2.5. *If a chain (p, P) is an absorbing chain with a unique absorbing state, conditions (2.3) and (2.19) are fulfilled, $\bar{\varphi}(l)$ is given by formula (2.21), and $h(l) \equiv h$, $a_l(x, y) \equiv a(x, y)$, $l = 1, 2, \dots, N$, $x \in \mathcal{X}$, then*

$$\begin{aligned} &E\bar{\varphi}(k) = \varphi(k), \quad E\widehat{J}_{2(N)} = J_{(N)}, \\ &E\widehat{J}^2_{2(N)} = \left(p^2/\tilde{p},\ E(\bar{\varphi}^{(N)})^2\right) = \left(\Psi^{(N)},\ h^2/\tilde{g}\right). \end{aligned}$$

Moreover, $E(\bar{\varphi}^{(k)})^2$ satisfies the relations

$$\begin{aligned} E(\bar{\varphi}(0))^2 &= h^2/\tilde{g}, \\ E(\bar{\varphi}(x; k+1))^2 &= \sum_y \frac{R^2(x, y)}{\tilde{P}_k(x, y)}\, E(\bar{\varphi}(x, k))^2 / (1 - g(x)) + \\ &\quad + h^2(x)/g(x), \end{aligned}$$

and $\Psi(N)$ is defined by the relations

$$\Psi(0) = p^2/\tilde{p},$$
$$\Psi(k+1) = (R_{sq}/\tilde{P}_{(k)})^T\,\Psi(k) + p^2/\tilde{p}, \quad k = 0, 1, \dots, N-1.$$

This theorem as does Theorem 2.1, enables us to find a fictitious chain for which an estimator has a minimal variance (equal to zero if $h(x) \geq 0$).

Consider a class of fictitious chains of the type

$$\tilde{p} = pq(0)/(p, q(0)),$$
$$\tilde{P}_k(x,y) = (1 - \tilde{g}(x,k))\; P_k(x,y)\, q(y;k) \Big/ \sum_{z\in\mathcal{X}'} P_k(x,z)\, q(z;k)\,, \tag{2.22}$$
$$k = 1, 2, \dots, N, \quad x, y \in \mathcal{X}'.$$

Summation of the second equality over y at $y \in \mathcal{X}'$ shows that $\tilde{g}(x;k) = 1 - \sum_{y\in\mathcal{X}'} \tilde{P}_k(x,y)$, i.e., $\tilde{g}(x;k)$ is the probability of jumping into $\{0\}$ a fictitious chain on the $(k+1)$-th step under the condition of $\xi_k = x$.

For absorbing Markov chains with absorbtion state $\{0\}$ the following theorem is valid.

Theorem 2.6. *For fictitious chains defined in (2.22) and satisfying conditions (2.3), (2.20), the minimum of variance of estimator $\widehat{J}_{2(N)}$ is attained if and only if*

$$q(k) = C_k|\varphi(N-k)|, \quad k = 1, 2, \dots, N,$$
$$\tilde{g}(x;k) = h(x)/|\varphi(x;N-k)|, \quad x \in \mathcal{X}, \quad k = 1, 2, \dots, N,$$

where C_k are arbitrary positive constants, $\varphi(k) = E(\bar{\varphi}(k))$. Moreover,

$$\tilde{p} = p|\varphi(N)|/J_{(N)},$$
$$\tilde{P}_k(x,y) = P(x,y)\,|\varphi(y;N-k)|\,/\,|\varphi(x;N-k-1) - h(x;k)|,$$

$k = 1, 2, \dots, N$. *Moreover, if $h(x) \geq 0$ then $\widehat{J}_{2(N)} = J_{(N)}$ with probability equals to 1.*

The proofs of this theorem and of Theorem 2.2 are the same.

3. Functional J estimation

Let $\{\xi_k;\ k \geq 0\}$ be a discrete absorbing Markov chain with the states space $\mathcal{X}' = \{0\} \cup \mathcal{X}$, $\mathcal{X} = \{1, 2, \dots\}$ or $\mathcal{X} = \{1, 2, \dots, n\}$ and a unique absorbing state $\{0\}$. Then a transition matrix (denote it by P) has the form (see, e.g.,[55])

$$P = \begin{pmatrix} 1 & 0 & \dots & 0 \\ q & & Q & \end{pmatrix}.$$

As is known from [55], a matrix

$$W = \sum_{k=0}^{\infty} Q^k$$

has only finite elements, therefore there exists an inverse matrix to $I - Q$ and

$$(I - Q)^{-1} = W.$$

In the case of a countable infinite state space, suppose, additionally, that

$$\left((I - Q)^{-1}|h|\right)(x) < \infty, \quad x \in \mathcal{X}. \tag{3.1}$$

For the finite chain ($\mathcal{X}' = \{0, 1, \dots, n\}$), this inequality is derived from the finiteness of the values $h(x), \ x \in \mathcal{X}'$.

Consider the problem of the functional J estimation with $a_{ji} \equiv 1$

$$J = E \sum_{k=0}^{\infty} h(\xi_k) = \sum_{k=0}^{\infty} p^T Q^k h = p^T (I - Q)^{-1} h.$$

Functions $\varphi(x; k)$ defined in (2.5) have the form

$$\varphi(k) = (I + Q + \dots + Q^k) h.$$

Since the series

$$\sum Q^k h$$

converges absolutely by virtue of condition (3.1), then there exists the finite limit

$$\varphi = \lim_{k \to \infty} \varphi(k) = (I - Q)^{-1} h.$$

So the limit of the quantity $E\widehat{J}_{1(N)}$, where $\widehat{J}_{1(N)}$ is defined in (2.2), for $N \to \infty$ exists and equals

$$\lim_{N \to \infty} E\widehat{J}_{1(N)} = \lim_{N \to \infty} (p, \varphi(N)) = (p, \varphi) = J.$$

Likewise, we get that, under the condition that matrix $I - R$ is invertible,

$$\lim_{N \to \infty} E\widehat{J}_{1(N)} = \lim_{N \to \infty} (p, \varphi(N)) = (p, \varphi) = J,$$

where

$$\varphi(N) = (I + R + \dots + R^N) h,$$
$$J = p^T (I - R)^{-1} h.$$

Let a matrix $I - R_{sq}/\tilde{P}$ be invertible where $\tilde{P}$ is a transition matrix of a fictitious chain satisfying the conditions

$$\begin{array}{llll} \text{from} & P(x,y) > 0 & \text{it follows that} & \tilde{P}(x,y) > 0, \\ \text{from} & p(x) > 0 & \text{it follows that} & \tilde{p}(x) > 0, \end{array} \tag{3.2}$$

for $x, y \in \mathcal{X}'$.

Also, let the condition

$$\Big((p^2/\tilde{p})^T (I - R_{sq}/\tilde{P})^{-1} \Big)(x) < \infty, \quad x \in \mathcal{X}'. \tag{3.3}$$

be valid.

Then there exists the limit

$$\Psi^T = \lim_{N\to\infty} \Psi^T(N) = \lim_{N\to\infty} \sum_{k=0}^{N} (p^2/\tilde{p})^T (R_{sq}/\tilde{P})^k =$$
$$= (p^2/\tilde{p})^T (I - R_{sq}/\tilde{P})^{-1}.$$

While proving Theorem 2.1, it was verified that

$$E\widehat{J}^2_{1(N)} = \sum_{k=1}^{N} \left(l(k),\ 2\varphi(N-k)h - h^2 \right),$$

where $l^T(k) = l^T(0)(R_{sq}/\tilde{P})^k$. Hence,

$$\lim_{N\to\infty} E\widehat{J}^2_{1(N)} = (\Psi,\ 2\varphi h - h^2).$$

Since a chain with probability equal to 1 hits into an absorbing state for a finite number of steps, then

$$\widehat{J}_1 = \lim_{N\to\infty} \widehat{J}_{1(N)}$$

with probability equal to 1, where

$$\widehat{J}_1 = \sum_{k=0}^{\alpha} h(\xi_k), \tag{3.4}$$

$\alpha = \{k\,; \xi_l \neq 0,\ l < k,\ \xi_k = 0\}$.

Note that φ and Ψ are unique (in the class of functions restricted to finite sets) solutions of the equations

$$\varphi = R\varphi + h \tag{3.5}$$

and

$$\Psi^T = \Psi^T (R_{sq}/\tilde{P}) + (p^2/\tilde{p})^T, \tag{3.6}$$

respectively, by virtue of the assumptions made above.

So, the following theorem is valid (see also the similar result in [29]).

Theorem 3.1. *For a discrete absorbing Markov chain with state space $\mathcal{X}' = \{0\} \cup \mathcal{X}$ and unique absorbing state $\{0\}$, under conditions (3.1) – (3.3), the estimator $\widehat{J}_1$ defined in (3.4), is an unbiased estimator of the functional J and*

$$\mathcal{D}\widehat{J}_1 = (\Psi,\ 2\varphi h - h^2) - J^2,$$

where φ and Ψ are the iterative solutions of equations (3.4) and (3.5), respectively.

Similarly applying a limit transition to the results of Theorem 2.2, we have the following result.

Theorem 3.2. *If the conditions of Theorem 3.1 hold, then the minimum of $\mathcal{D}\widehat{J}_1$ in the class of estimators defined by fictitious chains of the type*

$$\tilde{p} = pq/(p,q),$$
$$\tilde{P}(x,y) = R(x,y)\,q(y) \Big/ \sum_{y\in\mathcal{X}} R(x,y)\,q(y),$$

is attained if and only if $q = c|\varphi|$, where c is an arbitrary positive number.
Besides

$$\tilde{p} = p|\varphi|/J,$$
$$\tilde{P}(x,y) = R(x,y)\,|\varphi(y)|\ /\ |\varphi(x) - h(x)|.$$

If $h(x) \geq 0$, then $\mathcal{D}\widehat{J}_1 = 0$ and the estimator $\widehat{J}_1$ is independent of the simulating chain path.

Define the estimator $\widehat{J}_2$ as follows

$$\widehat{J}_2 = \frac{p(\tilde{\xi}_0)}{\tilde{p}(\tilde{\xi}_0)}\, v(\tilde{\xi}_0,\tilde{\xi}_1)\dots v(\tilde{\xi}_{k-1},\tilde{\xi}_k)\, h(\tilde{\xi}_k)/\tilde{g}(\tilde{\xi}_k),$$

where k is the index of the last step before an absorption and $\widehat{J}_2 = 0$ if an absorption does not occur. Then, for every path $\xi_0, \xi_1, \dots$, a number N such that $\widehat{J}_2 = \widehat{J}_{2(N)}$, $N \geq k$ will exist if absorption occurs at the $k+1$th step and $\widehat{J}_2 = 0$ if an absorption does not occur. Therefore, if such a limit exists, $\widehat{J}_2 = \lim_{N\to\infty} \widehat{J}_{2(N)}$ with probability equal to 1. The existence of the limit, the nonbias of the estimator $\widehat{J}_2$, and the formula for its variance, are established by the same technique as for the estimator $\widehat{J}_1$. Hence, define net results only.

Let $\tilde{q}(x) = 1 - \sum_{y\in\mathcal{X}} \tilde{P}(x,y)$.

Theorem 3.3. *If the conditions of Theorem 3.1 are valid, then with probability equal to 1, there exists the limit*

$$\widehat{J}_2 = \lim_{N\to\infty} \widehat{J}_{2(N)}$$

and

$$E\widehat{J}_2 = J, \quad E\widehat{J}_2^2 = (\Psi,\ h^2/\tilde{g}).$$

Theorem 3.4. *If the conditions of Theorem 3.1 are valid, then the minimum $\mathcal{D}\widehat{J}_2$ in the class of estimators defined by fictitious chains of the type*

$$\tilde{p} = pq/(p,q),$$
$$\tilde{P}(x,y) = (1-\tilde{g}(x))\, R(x,y)\, q(y) \Big/ \sum_{z\in\mathcal{X}} R(x,z)\, q(z)\,,$$

is attained if and only if $q = c|\varphi|$, $\tilde{g} = h^2/|\varphi|$, where $\varphi = (I-R)^{-1}h$, c is an arbitrary positive number. Besides

$$\tilde{P}(x,y) = R(x,y)\,|\varphi(y)| \;/\; |\varphi(x)|.$$

If $h(x) \geq 0$, $x \in \mathcal{X}'$, then $\mathcal{D}\widehat{J}_2 = 0$.

4. Investigation of stationary behavior

Here we will consider finite regular Markov chains with a transition matrix P and a stationary distribution π.

First, let us consider the behavior of the estimators of functional $J_{(N)}$ given in Section 2. Such estimators divided by N for $N \to \infty$ are J_{st}'s consistent estimators only in the case of a fictitious chain and (p, P) equality. Therefore, specific approaches are required to increase the efficiency of the functional J_{st} estimator. Two of them are connected with the embedded chains and the regeneration techniques, respectively. Within the framework of these techniques, some results have been obtained in papers which are systematized and generalized in this chapter.

4.1. Investigation of estimators in Section 2.

Let $\widehat{J}_{1(N)}$ be defined by formula (2.1) or, which is the same, by formula (2.2) with $\tilde{p} = p$, $\tilde{P} = P$, i.e.

$$\widehat{J}_{1(N)} = \sum_{i=0}^{N} h(X_i), \tag{4.1}$$

where $X_0, X_1, \ldots, X_N$ is the chain (p, P) path.

Consider an estimator $\widetilde{J}_{1(N)} = \widehat{J}_{1(N)}/N$. Denote by ${}_lP$, a matrix obtained from P by l-th column units changing to zeros.

Lemma 4.1. *Let P be the transition matrix of a regular Markov chain with a finite state space. Then $\lambda_{\max}({}_lP) < 1$.*

Proof. Note that the transition matrix of a regular Markov chain with a finite state space cannot have zero columns. In fact, assume otherwise. Let l be an index of the zero column. Then $P^k = PP^{k-1}$ has the l-th zero column for any k. On the other hand, such N, that $(P^N)_{ij} > 0$ for any i, j, exists for any regular chain. This contradiction shows that a matrix P has no zero columns. Let now $\lambda = \lambda_{\max}({}_lP) \geq 1$. Then there exists a vector $\pi_{(l)}$ such that $\sum \pi_{i(l)} = 1$ and

$$\pi_{(l)}^T\, {}_lP = \lambda \pi_{(l)}^T.$$

Sum up the elements of the left and right sides. In the left side. we have

$$\sum_{ij} \pi_{i(l)}\, P_{ij(l)} = \sum_i \pi_{i(l)} \sum_j P_{ij(l)} < 1,$$

since $\sum_j P_{ij(l)} < \sum P_{ij}$ for some i.

In the right side, we have

$$\lambda \sum \pi_{i(l)} = \lambda \geq 1.$$

The obtained contradiction proves the lemma.

Let ${}_l\varphi$ be the solution of the equation ${}_l\varphi = {}_lP\ {}_l\varphi + \tilde{h}$, where $\tilde{h} = h - g\bar{h}$, $\bar{h} = (1, \dots, 1)^T$. As $\lambda_{\max}({}_lP) < 1$, then there exists a matrix inverse to $I - {}_lP$ which is denoted by ${}_lG$. The following result characterizes the properties of the estimator $\tilde{J}_{1(N)}$.

Theorem 4.1. *Let (p, P) be a regular Markov chain with the state space $\mathcal{X} = \{0, 1, \dots, n\}$. An estimator $\widehat{J}_{1(N)}$ is consistent, asymptotically unbiased and normal for any $l \in \mathcal{X}$, besides*

$$\lim D(\sqrt{N}\, \widehat{J}_{1(N)}) = \sum_{i=0}^{n} \pi_i (2\ {}_l\varphi_i \tilde{h}_i - \tilde{h}_i^2) = (\pi, \tilde{h}^2) + 2(\pi_0 \tilde{h}, {}_lG\tilde{h}).$$

The proof of the theorem is based on the technique suggested in [89, section 7.4] and the results of Section 2.

Let us represent $\widehat{J}_{1(N)} - J_{st}$ in the form

$$\sum_{k=0}^{N\Lambda(\alpha_1 - 1)} \tilde{h}(X_k) + \sum_{i=1}^{i(N)} \tilde{\zeta}(i) + \sum_{k=i(N)+1}^{N} \tilde{h}(X_k), \tag{4.2}$$

where α_1 is a time unit of the first hitting of the chain $\{X_n;\ n \geq 0\}$ path into state 0: $\alpha_1 = \inf\{n;\ X_n = 0\}$, $\alpha_1 \Lambda N = \min\{\alpha_1,\ N\}$,

$$\tilde{\zeta}(i) = \sum_{k=\alpha_i}^{\alpha_{i+1}-1} \tilde{h}(X_k).$$

α_i is a time unit of the i-th hitting into state 0, $\quad i(N) = \inf\{i;\ \alpha_{i+1} > N\}$.

Note that

$$E\alpha_1 = p^T \sum_{k=0}^{\infty} ({}_0P)^k e_0 = p^T (I - {}_0P)^{-1} e_0 < \infty,$$

as $\lambda_{\min}({}_0P) < 1$, where $e_0 = (1, 0, \dots, 0)^T$.

The first term in expansion (4.2) is restricted from above by the quantity $\sum_{k=0}^{\alpha_1 - 1} |\tilde{h}(X_k)|$ which does not depend on N and is finite with probability equal to 1 (it follows from $E\alpha_1 < \infty$ that the number of addends is finite with probability equal to 1) and, on division by $\sqrt{N}$, converges to zero at $N \to \infty$ with probability equal to 1.

Therefore, since the third term is bounded by the same quantity with probability equal to 1,

$$\left(\sqrt{N}\right)^{-1} \sum_{k=i(N)+1}^{N} \tilde{h}(X_n) \to 0$$

in probability at $N \to \infty$.

Let us now investigate the second term in the expansion (4.2). By virtue of the Markov property of a chain (p, P), quantities $\beta_i = \alpha_{i+1} - \alpha_i$ as well as quantities $\tilde{\zeta}(i)$ are independent random variables with the same distribution, and from the Strong Large Numbers Law

$$\lim_{N \to \infty} i(N)/N = \lim_{k \to \infty} k/\beta_k = 1/E\beta_1 = \pi_0$$

with probability equal to 1. Hence, for any given $\varepsilon > 0$, there exists an integer $N(\varepsilon)$ such that for $N \geq N(\varepsilon)$ with probability equal to $1 - \varepsilon$, the inequality

$$\underline{N} \leq i(N) \leq \bar{N}$$

is valid, where $\underline{N} = \lfloor (1-\varepsilon)\pi_0 N \rfloor$, $\bar{N} = \lfloor (1+\varepsilon)\pi_0 N \rfloor$, ($\lfloor a \rfloor$ is the integer part of a). It is evident here that

$$\left| \sum_{i=1}^{i(N)} \tilde{\zeta}(i) - \sum_{i=1}^{\lfloor \pi_0 N \rfloor} \tilde{\zeta}(i) \right| \leq 2 \sum_{i=\underline{N}}^{\bar{N}} |\tilde{\zeta}(i)|.$$

By virtue of Tchebyshev's inequality for any $c > 0$,

$$P\left\{ \sum_{i=1}^{\bar{N}} |\tilde{\zeta}(i)| \geq \sqrt{N}\, c \right\} \leq N^{-1} c^{-2} (\bar{N} - \underline{N}) \mathcal{D} \tilde{\varphi}(1) \leq$$

$$\leq 2 c^{-2} \varepsilon \pi_0 \mathcal{D} \tilde{\zeta}(1).$$

The finiteness for a quantity $\mathcal{D}\tilde{\zeta}(1)$ and an explicit form for it will be established below. By virtue of the theorem of monotone convergence, from this inequality it follows that

$$\left| N^{-1/2} \sum_{i=1}^{i(N)} \tilde{\zeta}(i) - N^{-1/2} \sum_{i=1}^{\lfloor \pi_0 N \rfloor} \tilde{\zeta}(i) \right| \to 0$$

in probability. Therefore, in probability, the equality

$$\lim_{N\to\infty} N^{-1/2}\sum_{k=0}^{N}\tilde{h}(X_k) = \lim_{N\to\infty} N^{-1/2}\sum_{i=1}^{i(N)}\tilde{\zeta}(i) =$$
$$= \lim_{N\to\infty} N^{-1/2}\sum_{i=1}^{\lfloor N\pi_0\rfloor}\tilde{\zeta}(i) = \sqrt{\pi_0}\lim_{m\to\infty} m^{-1/2}\sum_{i=1}^{m}\tilde{\zeta}(i)$$

takes place.

By the Central Limit Theorem, the last quantity has a distribution $N(0,\pi_0\mathcal{D}\tilde{\zeta}(1))$. Let $\zeta(i)=\sum_{k=\alpha_i}^{\alpha_{i+1}-1}h(X_k)$. Then

$$\tilde{\zeta}(i) = \zeta(i) - \sum_{k=\alpha_i}^{\alpha_{i+1}-1} J_{st}\bar{h}(X_k) = \zeta(i) - J_{st}\beta_i.$$

For $N\to\infty$,

$$\tilde{J}_{1(N)} = \widehat{J}_{1(N)}/N \to E\,\zeta(1)\pi_0$$

in probability, i.e., $\tilde{J}_{1(N)}$ is a consistent estimator of $\pi_0 E\zeta(1)$. It is evident that

$$\lim_{N\to\infty} E(N\tilde{J}^2_{1(N)}) = \pi_0\mathcal{D}\zeta(1).$$

Let us use Theorem 3.1 for a chain $(e_0, {}_0P)$ for the computation of $E\zeta(1)$ and $E\zeta^2(1)$. The estimator $\widehat{J}_1$ for this chain coincides with $\zeta(1)$. Hence, $E\zeta(1) = E\widehat{J}_1 = (\Psi, h)\pi_0$, $E\zeta^2(1) = E\widehat{J}_1^2 = (\Psi,\ 2\,{}_0\varphi h + h^2)$, where ${}_0\varphi$ is the solution of the equation ${}_0\varphi = {}_0P\,{}_0\varphi + h$.

It is easy to check that $\Psi^T = (1, \pi_1/\pi_0, \dots, \pi_n/\pi_0)$ satisfies the equation $\Psi^T = \Psi^T\,{}_0P + e_0$:

$$\Psi^T\,{}_0P = \Psi^T P - (\Psi^T e_0, 0, \dots, 0)^T = \pi^T P/\pi_0 - e_0^T = \Psi^T - e_0^T.$$

Hence it follows that

$$E\zeta(1) = (\Psi,\ h) = \sum_{i=0}^{n}\pi_j h_j/\pi_0, \quad E\tilde{\zeta}(1) = (\Psi,\ \tilde{h}) = 0,$$

$$\lim_{N\to\infty} E\tilde{J}_{1(N)} = \pi_0\zeta(1) = \sum_{i=0}^{n}\pi_j h_j = J_{st},$$

$$E\tilde{\zeta}^2(1) = \sum_{i=0}^{n}\Psi_i(2\,{}_0\varphi_i\tilde{h}_i - \tilde{h}_i^2) = \sum_{i=0}^{n}\pi_i/\pi_0(2\,{}_0\varphi_i\tilde{h}_i - \tilde{h}_i^2),$$

$$\lim_{N\to\infty}\mathcal{D}(\sqrt{N}\tilde{J}_{1(N)}) = \pi_0\mathcal{D}\zeta(1) = \sum_{i=0}^{n}\pi_i(2\,{}_0\varphi_i\tilde{h}_i - \tilde{h}_i^2).$$

Substituting the expression ${}_0\varphi = (I - {}_0P)^{-1}\tilde{h}$ for ${}_0\varphi$ and taking into account that ${}_0P(I - {}_0P)^{-1} = (I - {}_0P)^{-1} - I$, we have the formula

$$\lim_{N\to\infty} \mathcal{D}(\sqrt{N}\tilde{J}_N) = (\pi,\, \tilde{h}^2) + 2(\pi_0\tilde{\varphi},\, {}_0G\tilde{h}).$$

As in the definition of $\alpha_1, \alpha_2, \ldots$, one can choose any state $l \in \mathcal{X}$ instead of 0, then the theorem is complete (expressions for the limit variance $\sqrt{N}\tilde{J}_N$ evidently provide the same value for different l).

Denote by $\widehat{J}_{1(N)}(h_{(1)})$ and $\tilde{J}_{1(N)}(h_{(2)})$ quantities $\tilde{J}_{1(N)}$ for the functionals $h_{(1)}$ and $h_{(1)}$. Similarly to Theorem 4.1, the following result is obtained.

Theorem 4.2. *Let (p, P) be a regular Markov chain with state space $\mathcal{X} = \{0, 1, \ldots, n\}$. A joint distribution of the estimators $\tilde{J}_{1(N)}(h_{(1)})$ and $\tilde{J}_{1(N)}(h_{(2)})$ is asymptotically normal and the quantity*

$$\sum_{i=0}^{n} \pi_i \left({}_l\varphi_i(h_{(1)})\, h_{(2)i} + {}_l\varphi_i(h_{(2)}) h_{(1)i} - h_{(1)i} h_{(2)i} \right)$$

does not depend on l but is equal to $\lim_{N\to\infty} \sqrt{N}$ $\mathrm{Cov}\{\tilde{J}_{1(N)}(h_{(1)}),\ \tilde{J}_{1(N)}(h_{(2)})\}$, where ${}_l\varphi(h_{(k)}) = {}_lP\ {}_l\varphi(h_{(k)}) + h_{(k)},\ k = 1, 2$.

Note that the estimator $\tilde{J}_{1(N)}$ is nonconsistent for any fictitious chain.

Theorem 4.3. *For any regular Markov chain, the estimator $\tilde{J}_{1(N)} = \widehat{J}_{1(N)}/N$, where $\widehat{J}_{1(N)}$ is defined by formula (3.2), is nonconsistent for any choice of $(\tilde{p}, \tilde{P}) \neq (p, P)$.*

The proof of this theorem can be found in [29,74]. Fictitious chains can be used within the framework of the two approaches considered below.

4.2. Regeneration technique.

The idea of the regeneration technique goes back to Cox and Smith's work [20] and has been systematically developed in the works by Krein and Lemoin. This method for the estimation of the stationary distribution of finite Markov chains was investigated in [22]. Application of estimators with zero variance in terms of the results of Section 2 seems to be new.

For finite Markov chains, the technique is based on the fact that the segments of such chains embedded between two successive hits into any fixed state, are probabilistically equal to each other. In other words, any state is a state of regeneration. In fact, the idea of the regeneration method has been already applied for the proof of Theorem 4.1. Now let us consider estimators of a special type.

Consider an estimator of the type

$$\tilde{J}_{3(k)} = \frac{1}{k}\sum_{i=1}^{k} \zeta(i) \Big/ \frac{1}{k}\sum_{i=1}^{k} (\alpha_i - \alpha_{i-1}), \tag{4.3}$$

where $\alpha_0 = 0$, $X_0 = l$, $\alpha_1, \dots, \alpha_k$ are successive points of a chain (e_l, P) returning to state l.

The estimator $\tilde{J}_{3(k)}$ has the following advantages. First, its variance has a convenient formula in terms of functionals calculated in the simulation process. Second, the choice of the initial distribution and of the moment of discontinuance of the simulation has a simple and well-grounded solution. Determine the efficiency of the estimator $\tilde{J}_{1(N)}$ by the quantity $N\mathcal{D}\tilde{J}_{1(N)}$ and the efficiency of $\tilde{J}_{3(k)}$ by $E\alpha_1 \mathcal{D}\tilde{J}_{3(k)}$ proportional to the mean time of simulation under a fixed accuracy. The next theorem indicates that the efficiency of the estimator $\tilde{J}_{3(k)}$ is the same as that of $\tilde{J}_{1(N)}$ and does not depend on the choice of the regeneration state.

Theorem 4.4. *If $\{X_s;\ s \geq 0\}$ is a regular Markov chain with state space $\mathcal{X} = \{0, \dots, n\}$, then $\tilde{J}_{3(k)}$, defined by formula (4.3), is strongly consistent. The distribution of the quantity $\sqrt{k}(\tilde{J}_{3(k)} - J_{st})$ for $k \to \infty$ converges to a normal one with parameters $(0, \sigma_l^2)$, where the quantity $\sigma_l^2 E\alpha_1 = \sigma^2/\pi_l$ is independent of the choice of regeneration state and equals*

$$\left(\mathcal{D}\zeta(1) - 2J_{st}\operatorname{Cov}(\varphi(1), \alpha_1) + J^2\mathcal{D}\alpha_1\right) =$$

$$= \sum_{i=0}^{n} \pi_i (2\,{}_l\varphi_i \tilde{h}_i^2 - \tilde{h}_i^2) = \lim_{N \to \infty} \mathcal{D}(\sqrt{N}\tilde{J}_{1(N)}).$$

Proof. As the numerator and denominator in (4.3) are the sums of independent, equally distributed random variables and $E\alpha_1$ and $E\zeta(1)$ are finite, then by virtue of the Strong Large Numbers Law with probability 1,

$$\tilde{J}_{3(k)} \to E\zeta(1)/E\alpha_1 = J_{st},$$

i.e., $\tilde{J}_{3(k)}$ is a strongly consistent estimator J_{st}.

As $\frac{1}{k}\sum_{s=1}^{k} \beta_s \to E\alpha_1 = \frac{1}{\pi_l}$ with probability 1 and $\mathcal{D}\left(\frac{1}{k}\sum_{s=1}^{k}\beta_s\right) = (\mathcal{D}\alpha_1)/k \to 0$ for $k \to \infty$ then by virtue of Tchebyshev's inequality for any $\varepsilon > 0$, there exists such $k(\varepsilon)$ that for $k \geq k(\varepsilon)$ with probability equal to $1 - \varepsilon$

$$\frac{1}{\sqrt{k}} \pi_l (1 - \varepsilon) \left| \sum_{s=1}^{k} \tilde{\zeta}(s) \right| \leq \sqrt{k}\,|\tilde{J}_{3(k)} - J_{st}| \leq$$

$$\leq \frac{1}{\sqrt{k}} \pi_l (1 + \varepsilon) \left| \sum_{s=1}^{k} \tilde{\zeta}(s) \right|.$$

Moreover, $\tilde{\zeta}(l)$ are independent random variables with the same distribution and, by virtue of the Central Limit Theorem, $\sum_{j=1}^{k} \tilde{\zeta}(s)/k$ converges by distribution to a random variable with normal distribution with parameters $(0,\ \mathcal{D}\zeta(1))$. Hence, $\sqrt{k}\,(\tilde{J}_{3(k)} - J_{st})$ converges by distribution to a random variable with parameters $(0,\ \pi_l^2\mathcal{D}\zeta(1))$. Hence, it follows that

$$\sigma^2 = \lim_{k\to\infty} \mathcal{D}\{\sqrt{k}\,(\tilde{J}_{3(k)} - J_{st})\} = \pi_l\sigma_1^2 = \sigma_1^2/E\alpha_1,$$

where $\sigma_1^2 = \lim_{k\to\infty} \mathcal{D}\{\sqrt{N}\,(\tilde{J}_{1(N)} - J_{st})\} = \pi_l\mathcal{D}\tilde{\zeta}(1)$. Finally, we have

$$\mathcal{D}\tilde{\zeta}(1) = \mathcal{D}(\zeta(1) - J_{st}\alpha_1) = \mathcal{D}\zeta(1) - 2J_{st}\,\mathrm{Cov}(\alpha_1,\ \zeta(1)) + \mathcal{D}\alpha_1$$

and

$$(E\alpha_1)\sigma^2 = \sigma_1^2 = \mathcal{D}\tilde{\zeta}(1)/E\alpha_1,$$

that completes the proof of Theorem 4.4.

The purely algebraic proof of the following fact is worth considering. The fact is that the expression for the variance in Theorem 4.1 is independent of l. Consider a pair of conjugate equations

$$\begin{aligned} {}_l\varphi &= {}_lP\,{}_l\varphi + \tilde{h}, \\ {}_l\Psi^T &= {}_l\Psi^T{}_lP + e_l, e_l = (0,\ldots,0,1,0,\ldots,0). \end{aligned}$$

The second equation has the explicit solution

$${}_l\Psi^T = (\pi_0/\pi_l,\ldots,\pi_n/\pi_l),$$

which is checked by substitution. As

$$({}_l\varphi,\, e_l) = ({}_l\Psi,\, \tilde{h}) = \sum_{i=0}^{n} \pi_i h_i/\pi_l - J\sum_{i=0}^{n} \pi_i/\pi_l = 0,$$

then ${}_l\varphi_l = 0$.
Set

$$\begin{aligned} &Q_{(l)} = (P_{ij})_{i,j\neq 1},\ \ \tilde{h}_{(l)} = (\tilde{h}_1,\ldots,\tilde{h}_{l-1},\, \tilde{h}_{l+1},\ldots,\tilde{h}_n)^T, \\ &\varphi_{(l)} = ({}_l\varphi_1,\ldots,{}_l\varphi_{l-1},\, {}_l\varphi_{l+1},\ldots,{}_l\varphi_n)^T,\ \ \pi_{(l)} = (\pi_0,\ldots,\pi_{l-1},\, \pi_{l+1},\ldots,\pi_n)^T. \end{aligned}$$

From ${}_l\varphi_l = 0$, it follows that $\varphi_{(l)}$ is the solution of the equation $\varphi_{(l)} = Q_{(l)}\varphi_{(l)} + \tilde{h}_{(l)}$, whence

$$\varphi_{(l)} = (I - Q_{(l)})^{-1}\tilde{h}_{(i)}.$$

Now

$$({}_l\Psi\ {}_l\varphi\tilde{h}) = (\Psi_{(l)},\ \varphi_{(l)}\,\tilde{h}_{(l)}) = \frac{1}{\pi_0}\left(\pi_{(l)}\tilde{h}_{(l)},\ (I - Q_{(l)})^{-1}\tilde{h}_{(l)}\right).$$

Denote this expression by L. Note that

$$L = \det(I - Q_{(l)})_{\tilde{h}} \,/\, \det(I - Q_{(l)}),$$

where $(I - Q_{(l)})_{\tilde{h}}$ denotes a matrix $I - Q_{(l)}$ supplemented by the column $\tilde{h}_{(l)}$ in the first position, then by the line $(0, \pi_0 \tilde{h}_0, \dots, \pi_{l-1}\tilde{h}_{l-1}, \pi_{l+1}\tilde{h}_{l+1}, \dots, \pi_n \tilde{h}_n)$ in the first position:

$$(I - Q_{(l)})_{\tilde{h}} = \begin{pmatrix} 0, & \pi_0 \tilde{h}_0, & \dots, & \pi_{l-1}\tilde{h}_{l-1}, & \pi_{l+1}\tilde{h}_{l+1}, & \dots, & \pi_n \tilde{h}_n \\ \tilde{h}_{(l)} & & & I - Q_{(l)} & & & \end{pmatrix}.$$

It can be checked by expanding the determinant $(I - Q_{(l)})_{\tilde{h}}$ in the first line, and each of the minors in the first column.

Add the elements of all columns, except the first one, to the column with the first unit $\pi_s \tilde{h}_s$. Multiply the line with the first unit $\tilde{h}_s$ by π_s and add to it a line beginning with $\tilde{h}_0$ multiplied by $\pi_0, \dots$. By virtue of relations $\sum_{j=0}^{n} p_{ij} = 1$, $\sum_{i=0}^{n} \pi_i \tilde{h}_i = 0$, after the division of the obtained line by π_l, we will obtain a matrix $(I - Q_{(s)})_{\tilde{h}}$. Hence, $\pi_s \det(I - Q_{(l)})_{\tilde{h}} = \pi_l \det(I - Q_{(s)})_{\tilde{h}}$. The same converter applied to a matrix $I - Q_{(l)}$, gives $\pi_s \det(I - Q_{(l)}) = \pi_l \det(I - Q_{(s)})$. Hence,

$$L = \det(I - Q_{(l)}) s_{\tilde{h}} \,/\, \det(I - Q_{(l)}) = \det(I - Q_{(s)})_{\tilde{h}} \,/\, \det(I - Q_{(s)})$$

for any $s, l = 0, \dots, n$.

Consider the problem of the application of fictitious chains within the framework of the regeneration approach. As estimators can be represented as sums of independent random variables with the same distribution but correlated in pairs($\zeta(k)$ and β_k), we are unable to make their variance equal to zero. On the basis of Theorems 3.1 and 3.2, we can only make the variance of the unbiased estimators of J_{st}/π_l and $1/\pi_l$ separately equal to zero. Consider an analogue of Theorem 3.2. As a chain $(e_l, {}_lP)$ at each cycle is simulated, then it may be replaced by a fictitious chain of the form $(e_l, {}_l\tilde{P})$, where the ${}_l\tilde{P}$ elements have the form

$$\tilde{P}(x, y) = P(x, y)(r(x)/r(y)), \quad x \neq 1, \quad \tilde{P}(l, y) \equiv 0. \tag{4.4}$$

Choose r so that $r(l) = 1$, $r_{(l)}^T = (r_0, \dots, r_{l-1}, r_{l+1}, \dots, r_n)$ is the solution of the equation

$$r_{(l)} = Q_{(l)} r_{(l)} + h.$$

Set

$$Y = h(X_m)/r(X_m),$$

where $m + 1$ is the moment of the breaking off of a fictitious chain. From Theorem 3.2, it follows that $EY = J_{st}/\pi_l$, $\mathcal{D}Y = 0$ and the estimator does not depend on the chain path. Replacing h by $\tilde{h}$, we have an estimator of $1/\pi_l$ with zero variance.

Similarly, we may construct estimators J_{st}/π_l and $1/\pi_l$ with zero variance on the basis of an estimator $\widehat{J}_1$, but here the path length is infinite.

From a practical point of view, one may often be content with the above-mentioned approach. But it would be worthwhile studing the problem of maximization of simulation efficiency being measured by the product of $\mathcal{D}\tilde{J}_3 E\alpha_1$ when estimating one or several functionals of the form $\sum_{i=0}^{n} \pi_i h_i$.

First of all, it is necessary to get the form of the variance of the fictitious chain weighted estimator. Let a chain have the form (4.4) and the initial distribution e_0, $r(0) = 1$ and r satisfies the inequality

$$Pr \leq r \tag{4.5}$$

(otherwise (4.4) will not define a transition matrix). Let P be a transition matrix of the regular Markov chain with the stationary distribution π. Consider an estimator

$$\tilde{J}_{4(k)} = \left(\sum_{s=1}^{k} Y(s)/k\right) \Big/ \sum_{s=1}^{k} (m_s + 1)/k,$$

where $Y(s) = h(X_{m_s}^{(s)})/r(X_{m_s}^{(s)})$, $\{X_i^{(s)}, i \geq 0\}$ are paths of the chain $(e_l, {}_l\tilde{P})$ where units of a matrix ${}_l\tilde{P}$ are defined by equality (4.4) and, for any N, satisfy the continuity condition (2.4). Suppose that $r_i > 0$ if $\tilde{h}_i \neq 0$. Set $r_i^{-1}\tilde{h}_i^2 = 0$ if $\tilde{h}_i = 0$. The following theorem holds.

Theorem 4.5. *Under the conditions formulated above, the estimator $\tilde{J}_{4(k)}$ for $k \to \infty$ is strongly consistent. Distribution of the quantity $\sqrt{k}\,(\tilde{J}_{4(k)} - J_{st})$ converges for $k \to \infty$ to the normal distribution with parameters $(0, \sigma_l^2)$, where*

$$\sigma_l^2 = \frac{1}{\pi_l} \sum \pi_i r_i^{-1} \tilde{h}_i^2 / q_i,$$

$$E(m_s + 1) = \pi_l \Big/ \sum_{i=0}^{n} r_i \pi_i\,, \quad q_i = 1 - \sum_{j=0}^{n} p_{ij}\, r_j / r_i,$$

$$i \neq l, \quad q_l = r_l.$$

The proof of Theorem 4.5 is similar to that of Theorem 4.1.

Consider the efficiency maximization problem of the estimator $\tilde{J}_{4(k)}$ by r. It is required to find a minimum of

$$V = E(m_s + 1)\sigma_l^2 = \sum_{i=0}^{n} \pi_i r_i^{-1} (\tilde{h}_i^2 / q_i) \sum_{i=0}^{n} \pi_i r_i.$$

Denote $v_i = q_i r_i$. By the definition of g, we have $r = (I - {}_lP)^{-1} v$. Let $a_i = \pi^T (I - {}_lP)^{-1} e_i$. As $\pi_i > 0$ then $a_i > 0$.

Now v takes the form

$$v = \sum_{i=0}^{n} \pi_i (\tilde{h}_i^2 / v_i) \sum_{i=0}^{n} \pi_i v_i .$$

Differentiating this expression with respect to v_i as independent variables, we find that an unconditional minimum v with respect to $v \geq 0$ is attained for

$$v_i = |h_i| \sqrt{\pi_i} \,/\, \sqrt{a_i} .$$

Since $r - {}_l P r = v \geq 0$, then condition (4.3) is valid.

4.3. Method of embedded chains.

The idea of the application of embedded chains for constructing the value J_{st} estimators was developed by I.Yu. Linnik [74] and elaborated upon in the work by Yu.N. Kashtanov [51] without using the term itself. We will obtain some more considerable results in another way on the basis of Theorem 2.2.

Divide a chain path $\{\xi_l;\ l \geq 0\}$ into segments of a length to be a multiple to some number $k \geq 1$. Consider a sequence of homogeneous Markov chains $\{\xi_i^{(s)};\ i = 0, 1, \dots, k\}$, $s = 0, 1, \dots, m$ with initial states $\xi_0^{(s)} = \xi_{ks}$ and the same set of transition matrices $\tilde{P}_{(1)}, \dots, \tilde{P}_{(k)}$. We shall construct an estimator of the value J_{st} in the form

$$\tilde{J}_{3(m)} = \frac{1}{m} \sum_{s=0}^{m-1} b(\xi_{ks}),$$

where

$$b(\xi_{ks}) = \sum_{l=1}^{k} \frac{P(\xi_0^{(s)}, \xi_1^{(s)})}{\tilde{P}_{(1)}(\xi_0^{(s)}, \xi_1^{(s)})} \cdots \frac{P(\xi_{l-1}^{(s)}, \xi_l^{(s)})}{\tilde{P}_{(1)}(\xi_{l-1}^{(s)}, \xi_l^{(s)})} h(\xi_l^{(s)};\ l).$$

For chains $\{\xi_i^{(s)};\ i = 0, 1, \dots, k\}$, $s = 0, 1, \dots, m$, let the continuity condition (3.2) hold. Then the expression in the right side always makes sense.

Let $h(x, l) = h(x)\delta_{xl}$. Since $b(\xi_{ks})$ under fixed ξ_{ks} coincides with the estimator $\widehat{J}_{1(N)}$ from Section 2, then by Theorem 2.1, $b(\xi_{ks})$ is an unbiased estimator of the value $\varphi(\xi_{ks};\ k)$, where

$$\varphi(x; k) = \sum_{s=0}^{k} \sum_{y \in \mathcal{X}'} P^s(x, y)\, h(y; s) = \sum_{y \in \mathcal{X}'} P^k(x, y)\, h(y).$$

From the Strong Large Numbers Law for Markov chains, we receive that, with probability 1,

$$\tilde{J}_{3(m)} \xrightarrow[m \to \infty]{} \sum_{x \in \mathcal{X}'} \pi(x)\, E\left(b(\xi_{ks}) \mid \xi_{ks} = x\right) =$$

$$= \sum_{x \in \mathcal{X}'} \pi(x) \sum_{y \in \mathcal{X}'} P^k(x, y)\, h(y) = \sum_{x \in \mathcal{X}'} \pi(x)\, h(x) = J_{st},$$

i.e., $\tilde{J}_{3(m)}$ is a strongly consistent estimator J_{st}.

Investigate the variance of this estimator. By the definition

$$\mathcal{D}\tilde{J}_{3(m)} = E\left(\frac{1}{m}\sum_{s=0}^{m-1} b(\zeta_s)\right)^2 - \left(E\frac{1}{m}\sum_{s=0}^{m-1} b(\zeta_s)\right)^2, \quad \zeta_s = \varepsilon_{ks}.$$

Designate

$$R_m(v) = E\sum_{s=0}^{m-v-1} b(\zeta_s)\, b(\zeta_{(s+v)})\,\frac{1}{m-v} - \left(E\frac{1}{m}\sum_{s=0}^{m-1} b(\zeta_s)\right)^2,$$

where $v = 0, 1, \dots, m-1$. We have

$$\mathcal{D}\tilde{J}_{3(m)} = \frac{1}{m}\left\{R_n(0) + 2\sum_{v=1}^{m-1} R_m(v)\,\frac{m-v}{m}\right\}.$$

Note that $R_m(v)$ for $v \neq 0$ is independent of fictitious chains paths. Moreover,

$$E\frac{1}{m}\sum_{s=0}^{m-1} b(\zeta_s) = \frac{1}{m}\sum_{s=0}^{m-1}\sum_{y\in\mathcal{X}'} P^k(\xi_{ks}, y)\, h(y),$$

$$E\frac{1}{m}\sum_{s=0}^{m-v} b(\zeta_s)\, b(\zeta_{s+v}) =$$

$$= \frac{1}{m}\sum_{s=0}^{m-v}\sum_{y\in\mathcal{X}'} P^k(\xi_{ks}, y)\, h(y) \sum_{y\in\mathcal{X}'} P^k(\xi_{k(s+v)}, y)\, h(y),$$

where expectation is taken for a fixed path $\{\xi_l;\ l \geq 0\}$.

Hence, by the Strong Large Numbers Law for Markov chains, we have that a limit $R_m(v)$ for $m \to \infty$ and any fixed v exists and is equal to

$$R(v) = \sum_{x\in\mathcal{X}'} \pi(x)\, \varphi(x; k)\, \varphi(x; k(v+1)) - J_{st}^2, \quad v > 0.$$

The problem of $R(0)$ minimization was stated and solved in the work by I.Yu. Linnik [74]. It is equivalent to the minimization of the limit $\frac{1}{m}\mathcal{D}\tilde{J}_{3(m)}$ for $m \to \infty$, since $R(v)$, $v \geq 1$ does not depend on the choice of fictitious chains in the class of fictitious chains with transition matrices of the type

$$\tilde{P}_{(l)}(x, y) = P(x, y)\, q_{(l)}(y),$$

where $q_{(l)}(y)$ are arbitrary vectors such that, for $\tilde{P}_{(l)}(x,y)$, the continuity condition (2.3) is valid.

The indicated minimum is attained for

$$q_{(l)}(y) = \varphi(y; k-1), \quad y \in \mathcal{X}'$$

and equal to

$$R(0) = R^*(0) = \sum_{y\in\mathcal{X}'} \pi(y)\,\varphi^2(y;k) - J_{st}^2.$$

This result derives from the conducted reasoning and from Theorem 2.2 in a stronger form, namely, apart from the indicated earlier result (proved by I.Yu. Linnik), under the described choice $q_{(l)}$ an estimator $\tilde{J}_{3(m)}$ does not depend on a fictitious chains path and the minimum of a value $R_m(0)$ under fixed ξ_{ks} is attained for any fixed m and equals

$$\frac{1}{m}\sum_{s=0}^{m-1} \varphi^2(\xi_{ks};k) - \left(\frac{1}{m}\sum_{s=0}^{m-1}\varphi(\xi_{ks};k)\right)^2.$$

Hence the expression for $R(0)$ follows by means of a limiting transition on the basis of the Strong Large Numbers Law for Markov chains.

Similarly, the same results are provided for the case of $h(x,l) \equiv h(x), \quad l = 0,1,\ldots,k$. Here $\varphi(k)$ has the form

$$\varphi(x;k) = \sum_{l=1}^{k}\sum_{y\in\mathcal{X}'} P_{(l)}(x,y)\,h(y).$$

Note that for $k \to \infty$

$$m\mathcal{D}\tilde{g}_m \to 0.$$

Indeed, for finite regular Markov chains, the inequality

$$\left|\pi(x) - P^{(k)}(x,y)\right| < \gamma^k$$

is valid for some $0 < \gamma < 1$. Hence,

$$\sum_{l=1}^{\infty} R_l =$$

$$= \sum_{l=1}^{\infty}\sum_{y\in\mathcal{X}'}\pi(x)\sum_{y\in\mathcal{X}'}P^{(k)}(x,y)\pi(y)\sum_{y\in\mathcal{X}'}P^{(kl)}(x,y) - J_{st}^2 \le$$

$$\le \sum_{y\in\mathcal{X}'} h^2(y)\pi(y)\sum_{l=1}^{\infty}\gamma^{k(l+1)} = \sum_{y\in\mathcal{X}'}\pi(y)h^2(y)\frac{\gamma^{2k}}{1-\gamma^k} \to 0$$

for $k \to \infty$ and

$$R_0 \le \sum_{y \in \mathcal{X}'} \pi(y) h^2(y) \gamma^{2k} \to 0$$

Therefore for $k \to \infty$,

$$\lim_{m \to \infty} m \mathcal{D} \tilde{g}_m = R_0 + 2 \sum_{l}^{\infty} R_l \to 0.$$

Finally, we have the following result.

Theorem 4.6. *For an arbitrary finite regular Markov chain, the estimator*

$$\tilde{J}_{3(k)} = \frac{1}{m} \sum_{i=0}^{m-1} b(\zeta_i),$$

where $b(\zeta_i)$ is defined by formula (4.5), and the corresponding fictitious chains satisfy the continuity condition (2.3) for $N = k$, is a strongly consistent, asymptotically unbiased estimator of the value J_{st} for $h(l) \equiv h$ and $h(x; k) = h(x)\, \delta_{kl}$.

In the class of fictitious chains of the type (4.6), a minimum of the value $\frac{1}{m} \mathcal{D} \tilde{J}_{3(m)}$, for any fixed m, is attained if and only if

$$q_{(l)}(x) = \varphi(x; k - 1).$$

For $q_{(l)}(x) = \varphi(x; k-1)$, $l = 1, 2, \dots, k$, the limit $\frac{1}{m} \mathcal{D} \tilde{J}_{3(m)}$ for $m \to \infty$ is equal to

$$R_0^* + 2 \sum_{l=1}^{\infty} R_l,$$

where

$$R_l = \sum_{y \to \infty} \pi(y)\, \varphi(y; k)\, \varphi(y; k(l+1)) - J_{st}^2, \quad l \ge 1,$$

$$R_0^* = \sum_{y \to \infty} \pi(y)\, \varphi^2(y; k) - J_{st}^2.$$

5. Method of rotation sampling

The investigation of the rotation sampling technique for the simulation of discrete Markov chains has been carried out by George Fishman [41]. We will study its efficiency for the estimators of a linear functional of a discrete distribution. This result will be used in the next chapter.

Let $p = \{p_i\}_{i=0}^{n}$ be a discrete distribution, where $\sum p_i = 1$, p_i is the probability of hitting into the state with index i. An ordinary simulation method is as follows

$$\xi^{(l)} = i \;\; \text{for} \;\; \alpha_l \in [q_i, q_{i+1}), \;\; q_0 = 0,$$

where $q_{i+1} = \sum_{s=0}^{i} p_s$, α_l, $l = 1, 2, \dots$ are independent random values uniformly distributed at the segment $[0,1]$. Denote such a distribution by $U(0,1)$ and write it as $\alpha \sim U(0,1)$ to indicate the fact that α has the distribution $U(0,1)$.

Consider a parallel simulation of k distributions p on the basis of dependent random values (the rotation sampling technique):

$$\xi^{(l)}(s) = i \quad \text{for} \quad \alpha_s^{(l)} \in [q_i, q_{i+1}),$$

where $\alpha_s^{(l)} = \overline{(\alpha_l + s - 1)/k}$, $s = 1, 2, \dots, k$, $\{\alpha_l\}$ are independent random values with a general distribution $U(0,1)$, $\overline{(\alpha_l + s - 1)/k}$ is the fractional part of a number $(\alpha_l + s - 1)/k$.

Consider the estimators of a linear functional

$$R = \sum_{i=0}^{n} a_i p_i,$$

where a_i are given real numbers for the standard simulation scheme and for using the rotation sampling technique

$$\widehat{R}_k = \sum_{i=0}^{n} a_i \hat{p}_i, \quad R'_k = \sum_{i=0}^{n} a_i p'_i,$$

where $\hat{p}_i = \sum_{l=1}^{k} \chi\left(\xi^{(l)} = i\right)/k$,

$$p'_i = \sum_{s=1}^{k} \chi\left(\xi^{(1)}(s) = i\right) \Big/ k.$$

It is evident, that

$$\widehat{R}_k = \sum_{l=1}^{k} a_l \xi^{(l)}/k, \quad R'_k = \sum_{s=1}^{k} a_s \xi^{(1)}(s)/k.$$

Theorem 5.1. *When simulating an arbitrary discrete distribution*

$$(1) \quad ER'_k = E\widehat{R}_k = R,$$

$$(2) \quad \mathcal{D}\widehat{R}_k = O(n/k), \quad \mathcal{D}R'_k = O(n/k^2) \quad \textit{for} \quad k \to \infty.$$

Proof. Direct computations give us the following results

$$E\widehat{R}_k = R;$$

$$E\hat{p}_i\hat{p}_j = E\sum_x^k \left(l = 1\xi^{(l)} = i\right) \sum_{l=1}^k x\left(\xi^{(l)} = j\right)/k^2 =$$

$$= \begin{cases} p_i p_j(1 - \frac{1}{k}) & \text{for} \quad i \neq j, \\ p_i^2(1 - \frac{1}{k}) + p_i/k, & i = j; \end{cases}$$

$$E\widehat{R}_k^2 = \sum_{i,j=0}^k a_i a_j E\hat{p}_i\hat{p}_j = \sum a_i a_j p_{ij}\left(1 - \frac{1}{k}\right) +$$

$$+\sum a_i^2 p_i^2\left(1 - \frac{1}{k}\right) + \sum a_i^2 p_i/k = R^2\left(1 - \frac{1}{k}\right) + \sum a_i^2 p_i/k;$$

$$\mathcal{D}\widehat{R}_k = \left(\sum a_i^2 p_i - \left(\sum a_i p_i\right)^2\right)\Big/ k = O(n/k) \quad (k \to \infty).$$

Note, that

$$p_i' = \frac{1}{k}\sum_{l=1}^k \chi_{[q_i, q_{i+1})}\overline{(\alpha + (l-1)/k)} = (\lfloor kp_i \rfloor + \gamma)/k,$$

where $\gamma = 1$ with probability $\overline{kp_i}$ and $\gamma = 0$ with probability $1 - \overline{kp_i}$.

Hence

$$Ep_i' = p_i \quad \text{and} \quad ER_k' = R;$$

$$\mathcal{D}p_k' = \overline{kp_i}(1 - \overline{kp_i})/k^2 \leq 1/4\,k^2,$$

$$\mathcal{D}R_k' = O(n/k^2) \quad (k \to \infty).$$

Theorem 5.1 is proved.

CHAPTER 3. PATH BRANCHING TECHNIQUE FOR DISCRETE MARKOV CHAINS

The idea of path branching technique was formulated by von Neumann under the title splitting and Russian roulette (see [49]). This approach was investigated in papers [59,86,90] but they do not contain a systematic theory. This chapter dwells on the construction of such a theory for estimating the characteristics of Markov chains with discrete state spaces and is based on the authors' papers [33–35,80–84]. Such chains are important in queueing theory and in a diversity of other fields. The results for discrete chains are more clear and advanced than for general Markov chains and therefore deserve separate consideration. At the same time, many results are readily expanded to the case of general chains to which the following chapter is devoted.

In Section 1, the general circuit branching technique for random processes having an embedded Markov chain is considered. In Sections 2–4 representations for variance of an estimator of the linear functional from a finite segment of a Markov chain and also from absorbing and regular chain paths are described. In Sections 5–8, a concept of the design of branching paths and various optimal criteria of this design are defined. Explicit representations for optimum designs are described. These representations enable us to construct designs without additional information, but on the basis of current results of simulation only. Also, a numerical procedure of optimal (in the sense of an ellipsoid of the parameters of an estimator's concentration) design construction is recorded. In Section 9, a chain with state space cardinality equal to 2 is studied. It is shown that the maximal efficiency defined as the ratio of the amount of expenditures for a simulation experiment without branching to that for an optimum experiment with branching does not exceed 2. Section 10 is devoted to the investigation of waiting and random walk processes for which asymptotically optimal designs have been already constructed. The efficiency in these designs proves to be considerable in the evaluation of the probability of rare events. In the last Section, extra results for a special type of chains which present interest when investigating the reliability of complex systems, have been obtained.

1. The branching technique – general outline

Let there be a computing algorithm that enables us to receive process $\{X_t\}$ paths. This process describes the behavior of a complex system where t may take either discrete values or values from an interval on the real axis. $t_0 = 0$ will be considered to be the initial point of system development and for any fixed t, the random variable X_t takes values in the set $\mathcal{X}$. Let $\mathcal{X}$ be represented in the form of a union of mutually disjoint sets: $\mathcal{X} = \cup_{i=0}^{n} \mathcal{X}_i$, $\mathcal{X}_i \cap \mathcal{X}_j = \emptyset$ for $i \neq j$. Denote by $t_1, t_2, \dots, t_k, \dots$ instants when the process $\{X_t\}$ goes over from the set $\mathcal{X}_{i_{k-1}}$ to $\mathcal{X}_{i_k}$. Make up a random process with a discrete time and a finite state space $\xi_k = i_k$ (i_k takes its value in the set $\{0, 1, \dots, n\}$). Let an estimate of the quantity

$$J = E\,\frac{1}{T}\int_0^T h(X_t)\,dt$$

or

$$J = E\frac{1}{N}\sum_{t=0}^{N} h(X_t),$$

be the purpose of simulation where h is a given measurable function. Let $1 = \beta_0 \leq \beta_1 \leq \ldots \leq \beta_n < \infty$, β_{i+1}/β_i be integers. The branching technique can be described as follows. For instances t_n, $\beta_{i_n}/\beta_{i_{n-1}} - 1$ doubles of the process $\{X_t\}$ path are formed at $i_n > i_{n-1}$. These doubles should be simulated independently of each other and of the $\{X_t\}$ path which has generated them, and if $i_n \leq i_{n-1}$, then the path simulation discontinues with probability $1 - \beta_{i_n}/\beta_{i_{n-1}}$ and continues with probability $\beta_{i_n}/\beta_{i_{n-1}}$.

The quantity

$$\tilde{J} = \sum_{i=0}^{n} \beta_i^{-1}(1/T)\int_0^T \sum_{\alpha=1}^{N_t} h\left(X_t^{(\alpha)}\right) \mathcal{X}\left(X_t^{(\alpha)} \in \mathcal{X}\right) dt,$$

where $\{X_t^{(\alpha)}\}$ is one of doubles a process $\{X_t\}$; N_t is their number at an instant t; $\mathcal{X}(.)$ is the corresponding event indicator, is, evidently, an unbiased estimator of the quantity J. Quantity $\tilde{J}$ variance depends on the technique parameters $\beta_1, \beta_2, \ldots, \beta_n$, and is to be investigated.

Technique modification is of special interest. In this modification, instead of paths modelling stopping with a given probability, paths modelling stops in accordance with their labels are introduced. The index of the process $\{\xi_k\}$ state, wherein a path arises, is used as a label, and modelling stopping is performed if a current index of process $\{\xi_k\}$ state is less than the value of the label.

Note that the decomposition and path cancellation procedure depends only on the process $\{\xi_k\}$ though $\{X_t\}$ paths can be much more complicated than those of $\{\xi_k\}$.

For many problems of queueing theory, the process $\{\xi_k\}$ is a Markov chain. Besides, $\{\xi_k\}$ is connected with the consolidation of random process states and has been studied in many papers devoted to the consolidation of the states of random processes [70]. It turns out that with "natural" suppositions on the sequence of processes $\{X_{t,1}\}, \{X_{t,2}\}, \ldots$, a process $\{\xi_{k,l}\}$ produced by $\{X_{t,l}\}$ for large l is "close"to a Markov chain. One intuitive sense of these suppositions is that the behaviour of process $\{X_{t,l}\}$ after hitting $\mathcal{X}_i$, weakly depends on the concrete state from which a motion of the process inside $\mathcal{X}_i$ starts, and this dependence weakens for $l \to \infty$.

The pointed-out concepts show that Markov chain simulation techniques, which are developed further, can be applied for many problems of complex systems simulation.

2. Estimation of a linear functional in the transient regime

Let $\{\xi_k;\ k \geq 0\}$ be a Markov chain with a discrete state space $\mathcal{X} = \{1, 2, \ldots, n\}$ or $\mathcal{X} = \{1, 2, \ldots\}$, $Q_{(k)}$ be a substochastic transition matrix at the k-th step,

$$Q_{(k)} = \{Q_{(k)}(x,y)\}_{x,y\in\mathcal{X}}, \quad Q_{(k)}(x,y) = P\{\xi_k = y \mid \xi_{k-1} = x\}, \quad k = 1, 2, \ldots, N,$$

ν be an initial distribution, $\nu = \{\nu(x)\}_{x\in\mathcal{X}}$, $\nu(x) = P\{\xi_0 = x\}$. Designate $Q(k) = Q_{(1)}\cdots Q_{(k)}$, ν is a vector column and $\nu^T Q$ is the product of a vector and matrix.

Let us consider the problem of estimation of the functional

$$J(h;N) = E\sum_{k=0}^{N} h(\xi_k;\, k) = \sum_{k=0}^{N} \nu^T Q^{(k)} h(k),$$

where $h(k) = \{h(x;k)\}$, $x \in \mathcal{X}$ is considered as a vector column, and $h(x;k)$ is an arbitrary real function given on $\mathcal{X}$ under a fixed $k = 0,1,2,\dots,N$.

Let $\beta(k) = \{\beta(x;k)\}_{x\in\mathcal{X}}$, $k = 0,1,2,\dots,N$ be a vector column with given positive coefficients. Determine the following parallel simulation procedure for independent copies of the chain $\{\xi_k;\, k \geq 0\}$, the number of which is explicitly regulated. Let an amount of (η_k) independent copies of the chain $\{\xi_k;\, k \geq 0\}$ be simulated at the k-th step, the computing rule η_k will be defined further. Denote by $v(x,y) = v_k(x,y)$ the number of copies realizing the jumping from x to y at the k-th step

$$\xi_{k-1}^{(i)} = x, \quad \xi_k^{(i)} = y, \quad i = i_1, \dots, i_{v(x,y)},$$
$$\sum_{x,y} v_k(x,y) = \eta_k .$$

Let $\lfloor a \rfloor$ denote a whole and $\bar{a}$ a fractional part of the real number a, $r_i(x,y) = 0$, $i \neq i_1,\dots,i_{v(x,y)}$. For the chain with the index of i_l $(l = 1,2,\dots,v(x,y))$, define a number r_{i_l} by the formula

$$r_{i_l}(x,y) = r_{i_l(k)}(x,y) = \begin{cases} \lfloor\beta(y;x)/\beta(x;k-1)\rfloor + 1, & \text{if } \zeta_{i_l} < \overline{\beta(y;k)/\beta(x;k-1)}, \\ \lfloor\beta(y;x)/\beta(x;k-1)\rfloor, & \text{otherwise,} \end{cases} \tag{2.1}$$

where ζ_{i_l} is a random value defined by one of the two following formulas

$$\zeta_{i_l} = \alpha_l, \tag{2.2}$$

where $\alpha_1,\dots,\alpha_{v(x,y)}$ are independent random values with uniform distribution on the segment $[0,1]$,

$$\zeta_{i_l} = \overline{\alpha + i_l/v(x,y)}, \tag{2.3}$$

where α is a random value with uniform distribution on the segment $[0,1]$.

Let $\{0\}$ be a fictitious state of a chain, $\beta(x;0) \equiv 1$. Let us suppose that the chain jumps from the state $\{0\}$ to $x \in \mathcal{X}$ at the zero step with probability $\nu(x)$. At the zero step, η_0 independent copies of the chain $\{\xi_k;\, k \geq 0\}$ are simulated. At the k-th step $(k = 1,2,\dots,N-1)$ i_l-th simulated copy jumping from the state x to y is replaced with

$r_{i_l(k)}(x,y)$ copies starting from y and further simulated independently. The total number of copies simulated at a $(k+1)$-th step equals to

$$\eta_{k+1} = \sum_{x,y} \sum_{i=1}^{\eta_k} r_i(x,y) = \sum_{x,y} \sum_{l=1}^{v_k(x,y)} r_{i_l(k)}(x,y), \tag{2.4}$$

$k = 0, 1, \ldots, N-2$. At the N-th step, copies are created but not simulated because the $(N+1)$-th step is absent.

Define an estimate of the functional $J(h, N)$ by the following formula

$$\widehat{J}_\beta(h; N) = (1/L) \sum_{k=0}^{N} \sum_{i=1}^{\eta_k} h(\xi_k^{(i)}; k)/\beta(\xi_k^{(i)}; k), \tag{2.5}$$

where $\xi_k^{(i)}$ $(i = 1, 2, \ldots, \eta_k)$ are values which are taken by copies of the chain $\{\xi_k;\ k \geq 0\}$ at the k-th step (they are denoted in the same way as the corresponding random variables, it is a standard approach which must not lead to ambiguity).

Let

$$\begin{aligned} &\varphi(x; 0) = h(x; N), \\ &\varphi(x; k+1) = \sum_{y \in \mathcal{X}} Q_{(k)}(x; y)\, \varphi(y; k) + h(x; N-k-1), \quad k = 0, 1, \ldots, N-1, \end{aligned} \tag{2.6}$$

like in Section 2.2

Note that

$$\varphi(x; N) = \sum_{k=1}^{N} \sum_{y} Q^{(k)}(x, y)\, h(y; k) + h(x; 0).$$

Introduce the notation

$$\mathcal{D}_Q\, \varphi(x; N-k) = \sum_{y \in \mathcal{X}} \varphi^2(y; N-k) - \left(\sum_{y \in \mathcal{X}} Q^{(k)}(x, y)\, \varphi(y; N-k) \right)^2, \quad x \in \mathcal{X},$$

$$\mathcal{D}_\nu \varphi(0; N) = \sum_{y \in \mathcal{X}} \nu(y)\, \varphi^2(y; N) - \left(\sum_{y \in \mathcal{X}} \nu(y)\, \varphi(y; N) \right)^2,$$

$$r_p(x; N-k) = \sum_{y \in \mathcal{X}} \gamma_k(x, y)\, Q_k(x, y) \left(\varphi(y; N-k) \frac{\beta(x; k)}{\beta(y; k-1)} \right)^2, \quad x \in \mathcal{X},$$

$$r_\nu(0; N) = \sum_{y \in \mathcal{X}} \nu(y)\, \gamma(y)\, (\varphi(y; N))^2,$$

where

$$\gamma_k(x,y) = \mathcal{D}\left(r_{i_1(k)}(x,y)\right), \quad \gamma(y) = \mathcal{D}\left(r_{i_1(0)}(x,y)\right), \quad k = 1,2,\dots,N.$$

When using formula (2.2), we have

$$\begin{aligned} &\gamma_k(x,y) = \bar{d}_k - \bar{d}_k^2, \quad d_k = \beta(y;k)/\beta(x;k-1), \quad k = 1,\dots,N-1, \\ &\gamma(y) = \bar{d} - \bar{d}^2, \quad d = \beta(y;0). \end{aligned} \tag{2.7}$$

Denote by $(\nu^T Q)(y)$ the vector component $\nu^T Q$ with index y and by $\widehat{T}_\beta$, the total number of simulated chain values (the total number of steps for all the copies)

$$\widehat{T}_\beta = \sum_{k=0}^{N-1} \eta_k .$$

The following result shows that $\widehat{J}_\beta(h;N)$ is an unbiased estimate of the functional $J(h;N)$ and gives a convenient representation for its variance.

In the case of a countable state space, suppose that

$$\nu^T|h|(0) < \infty, \quad \nu^T Q\,|h|(k) < \infty, \tag{2.8}$$

$k = 1,2,\dots,N-1$. For a finite state space, this inequality is trivial.

Let a branching process be given by formulas (2.1) and (2.2). Then the following theorem is valid.

Theorem 2.1. *For an arbitrary discrete Markov chain satisfying condition (2.8) for any positive $\beta(x;k)$, $k = 0,1,\dots,N$, $x \in \mathcal{X}$, the quantity $\widehat{J}_\beta(h;N)$ defined by formula (2.5) is an unbiased estimate of the functional $J(h;N)$. Besides*

$$\begin{aligned} \mathcal{D}\widehat{J}_\beta(h;N) = (1/L) \sum_{k=0}^{N-1} \sum_{y\in\mathcal{X}} \left(\nu^T Q^{(k)}\right)(y)\, \beta^{-1}(y;k) \left[\mathcal{D}_Q\, \varphi(y;N-k) + r_Q(y;N-k)\right] + \\ + (1/L)\,[\mathcal{D}_\nu \varphi(0;N) + r_\nu(0;N)]\, \beta^{-1}(0;0), \end{aligned}$$

$$E\widehat{T}_\beta = L\left(\sum_{k=1}^{N-1} \nu^T Q^{(k)} \beta(k) + 1\right).$$

Proof. First find the formula for $E\widehat{T}_\beta$. Note that for $r_{i_l(k)}(x,y)$ defined by (2.1), (2.2), the relations

$$\begin{aligned} Er_{i_l(k)}(x,y) &= \beta(y;k)/\beta(x;k-1), \\ \mathcal{D}r_{i_l(k)}(x,y) &= \bar{d}_k - \bar{d}_k^2, \end{aligned}$$

where $\bar{d}_k$ is defined in (2.7), are valid.

Direct calculation gives

$$Ev_0(0,y) = L\nu(y),$$
$$Ev_1(x,y) = L\nu(x)\,Q_{(1)}(x,y)\,\beta(x;0) = Ev_0(0,x)\,Q_{(1)}(x,y)\,\beta(x;0),$$
$$Ev_k(x,y) = \sum_z Ev_{k-1}(z,x)\,Q_{(k)}(x,y)\,\frac{\beta(x;k-1)}{\beta(z;k-2)}, \quad k = 2,\dots,N.$$

Hence it follows that

$$Ev_k(x,y) = L\left(\nu^T Q^{(k-1)}\right)(x)\,Q_{(k)}(x,y)\,\beta(x;k-1), \quad k \geq 1.$$

Using (2.4) and Wald's lemma on the independent random variables sum, we have the equality

$$E\eta_k = L\sum_{x,y} Ev_k(x,y)\,Er_{i_1(k)}(x,y) =$$
$$= L\sum_{x,y}\left(\nu^T Q^{(k-1)}\right)(x)\,Q_{(k)}(x,y)\,\beta(y;k) = L\nu^T Q^{(k)}\beta(k), \quad k \geq 1.$$

Hence

$$E\widehat{T}_\beta = \sum_{k=0}^{N-1} E\eta_k = L\sum_{k=1}^{N-1} \nu^T Q^{(k)}\beta(k).$$

Let $L = 1$. Evaluate the quantity

$$\bar{\varphi}_\beta(x;N-k) = \sum_{i=k}^{N}\sum_{\gamma=1}^{\eta_i} h\left(\xi_i^{(\gamma)};i\right)\Big/\,\beta\left(\xi_i^{(\gamma)};i\right)$$

under the condition that $\eta_k = 1$, $\xi_k = x$ and that simulation starts from the $(k+1)$-th step.

Calculate the expectation of the quantity $\bar{\varphi}_\beta(x;N-k)$. By virtue of the definition of $\bar{\varphi}_\beta(x;N-k)$

$$\bar{\varphi}_\beta(x;N-k-1) = h(x;k)/\beta(x;k) + \sum_y \zeta_k(x,y)\sum_{i=1}^{v(x,y)} \bar{\varphi}_\beta^{(i)}(y;N-k), \qquad (2.9)$$

where $\{\bar{\varphi}_\beta^{(i)}(y;N-k)\}$ are independent random variables with distribution coinciding with $\bar{\varphi}_\beta(y;N-k)$ and $\zeta_k(x,y)$ is a random value reflecting the transition of the chain $(\xi_k;\ k \geq 0)$ from x to y at the k-th step:

$$\zeta_k(x,y) = \begin{cases} 1, & \text{with probability } Q_{(k)}(x,y), \\ 0, & \text{with probability } 1 - Q_{(k)}(x,y). \end{cases}$$

We have

$$E\zeta_k(x,y) = Q_{(k)}(x,y), \quad E\zeta_k^2(x,y) = Q_k(x,y),$$
$$E\zeta_k(x,y)\,\zeta_k(x,z) = 0, \quad y \neq z.$$

Calculating expectation from both sides of (2.9), according to Wald's lemma, we have

$$E\bar{\varphi}_\beta(x;N-k-1) = h(x,k)/\beta(x;k) + \sum_y v(x,y)\,E\zeta_k(x,y)\,E\bar{\varphi}_\beta(y;N-k) =$$

$$= h(x;k)/\beta(x;k) + \sum_y \frac{\beta(y;k+1)}{\beta(x;k)}\,Q_{(k)}(x,y)\,E\bar{\varphi}_\beta(y;N-k).$$

Moreover,

$$E\bar{\varphi}_\beta(x;0) = h(x;N)/\beta(x;N).$$

Comparing these expressions to (2.6), we have

$$E\bar{\varphi}_\beta(y;N-k) = \varphi(x;N-k)/\beta(x;k).$$

For $L=1$, by virtue of the definition of $\varphi_\beta(x;N)$, we have

$$\widehat{J}_\beta(h;N) = \sum_x \zeta(x) \sum_{i=1}^{r(0,x)} \varphi_\beta^{(i)}(x;N), \tag{2.10}$$

where $\zeta(x) = 1$ with probability $\nu(x)$ and 0 with probability $1-\nu(x)$. Hence,

$$E\widehat{J}_\beta(h;N) = \sum_x \nu(x)\,\beta(x;0)\,E\bar{\varphi}_\beta(x;0) = \sum_x \nu(x)\,\varphi(x;N) = J(h;N).$$

Calculating expectation from the squares of both sides of (2.9), we have

$$\begin{aligned}
&\mathcal{D}\left[\bar{\varphi}_\beta(x;N-k)/\beta(x;k)\right] = \\
&= \sum_y \frac{\beta(x;k+1)}{\beta(y;k)}\,Q_k(x,y)\,\mathcal{D}\left[\bar{\varphi}_\beta(x;N-k-1)/\beta(x;k+1)\right] + \\
&+ \mathcal{D}_p\varphi(x;N-k) + r_p(x;N-k).
\end{aligned}$$

Hence

$$\mathcal{D}\left[\bar{\varphi}_\beta(x;N)/\beta(x;0)\right] = \sum_{k=0}^{N-1} Q^{(k)}(x,y)\,\frac{\beta(x;0)}{\beta(y;k)}\left[\mathcal{D}_Q\,\varphi(x;N-k) + r_Q(x;N-k)\right].$$

Calculating expectation from the squares of both sides of (2.10), we have

$$E\widehat{J}^2_\beta(h;N) = \sum_x \nu(x)\,\beta(x;0)\,\mathcal{D}\varphi_\beta(x;N) +$$
$$+ \sum_x \nu(x)\,\mathcal{D}r(x;0)\,\left(E\bar{\varphi}_\beta(x;N)\right)^2 +$$
$$+ \sum_x \nu(x)\,\beta^2(x;0)\,\left(E\bar{\varphi}_\beta(x;N)\right)^2.$$

Thus, in the case of $L = 1$, it follows that

$$\mathcal{D}\widehat{J}_\beta(h;N) = \sum_{k=1}^{N-1} (\nu^T Q^k)(y)\,\beta^{-1}(y;k)\,[\mathcal{D}_p\varphi(y;N-k) + r_p(y;N-k)] +$$
$$+ \mathcal{D}_p\varphi(0;N) + r_p(0;N).$$

In the case of $L > 1$, by virtue of the independence of the paths the right side is to be divided by L, which completes the proof of Theorem 2.1.

Note 2.1. Let $r_{i_l}(x,y)$ satisfy equalities (2.1) and (2.3). Suppose that the number of various values $\beta(x;k)$ is finite for all $k = 0, 1, \dots, N$ and does not exceed n'. Then the reasoning of Theorem 2.1 is valid, but the values $\gamma_k(x,y)$ and $\gamma(y)$ take the form

$$\gamma_k(x,y) = \mathcal{D} \sum_{l=1}^{\nu_k(x,y)} r_{i_l}(x,y) = O(n'/L),$$
$$\gamma(x) = \mathcal{D} \sum_{l=1}^{\nu(y)} r(y) = O(n'/L).$$

As Theorem 5.1 in chapter 2 is correct for the values ζ_{i_l} defined by (2.3), then

$$\begin{aligned} r_Q(x;N-k) &= O(n'/L), \\ r_p(0;N) &= O(n'/L). \end{aligned} \tag{2.11}$$

Note 2.2. Let us use formulas (2.1) and (2.2) to evaluate $r_{i_l}(x,y)$ as in Theorem 2.1. Assume that the quantities $\beta(x;k)/\beta(y;k-1)$ for any x, y are either integers or less than a unit. Denote by $B_{k,y}$ the set

$$B_{k,y} = \{x;\ x \in \mathcal{X},\ \beta(x;k)/\beta(y;k-1) \le 1\}.$$

Then the values $r_p(y;N-k)$, $r_p(0;N)$ take the form

$$(r_p(y;N-k) = \sum_{x \in B_{k,y}} Q_{(k)}(y,x)\,\varphi^2(x;N-k)\left(\frac{\beta(y;k-1)}{\beta(x;k)} - 1\right),$$
$$r_p(0;N) = 0$$

Hence,

$$\mathcal{D}\widehat{J}_{\beta}(h;N) =$$

$$= (1/L)\sum_{k=1}^{N-1}\left(\nu^T Q^{(k-1)}\right)(y)\,\beta^{-1}(y;k)\left[\mathcal{D}_p\varphi(y;N-k) - \sum_{x\in B_{k,y}}\varphi^2(x;N-k)\right] +$$

$$+ (1/L)\sum_{k=1}^{N-1}\sum_{x\in B_{k,y}}\left(\nu^T Q^{(k-1)}\right)(x)\,Q_{(k)}(x,y)\,\beta(y;k)\,\varphi^2(y;N-k)+$$

$$+ (1/L)\,\beta^{-1}(0;0)\,\mathcal{D}_p\varphi(0;N).$$

3. Absorbing chains

Let us consider homogeneous Markov chains $\{\xi_k;\ k\geq 0\}$ with a discrete state space $\mathcal{X} = \{1,2,\dots,n\}$ or $\mathcal{X} = \{1,2,\dots\}$. With the notation of Section 2, homogeneity means that $Q_{(k)} = Q,\ \ k = 1,2,\dots,\ \ Q^{(k)} = Q^k$. Denote

$$g(x) = 1 - \sum_y P(x,y).$$

By definition, $g(x) \geq 0$. Suppose that $g(x) > 0$, at least for one value $x \in \mathcal{X}$. Then, as is known from [55], a chain $\{\xi_k;\ k\geq 0\}$ is stopped with probability 1 and the inverse to the $I - Q$ matrix exists. Denote it by $W = (I-Q)^{-1}$. Add the fictitious state $\{0\}$: $\mathcal{X}' = \{0\}\cup\mathcal{X}$ to the considered chain and set

$$P\{\xi_{k+1} = 0 \mid \xi_k = x\} = g(x),$$
$$P\{\xi_{k+1} = 0 \mid \xi_k = 0\} = 1,\quad k = 0,1,\dots$$

A chain extended in this way is an absorbing chain; after hitting the state $\{0\}$, the chain remains in this state with probability equal to 1.

Also introduce a companion chain, the transition probability of which is defined by the matrix

$$P = \begin{pmatrix} p(0) & p^T \\ g & Q \end{pmatrix},$$

where $p = \{p(x)\}_{x\in\mathcal{X}}$ is some vector with nonnegative components and $\sum_{x\in\mathcal{X}'} p(x) = 1$. Suppose that there exists a unique vector $\pi = \{\pi(x)\}_{x\in\mathcal{X}'}$ such that $\sum \pi(x) = 1$ and

$$\pi^T P = \pi^T.$$

Consider the linear functional evaluation problem

$$J(h) = \sum_{k=0}^{\infty}\nu^T Q^{(k)} h = \nu^T (I-Q)^{-1} h = \nu^T W h,$$

where ν is the initial distribution and $h = \{h(x)\}$ is an arbitrary real vector such that

$$J(|h|) < \infty, \quad |h|(x) = |h(x)|, \quad x \in \mathcal{X}.$$

Let $L = 1$ and a branching process from Section 2 is simulated until all the chains $\{\xi_k^{(i)};\ k \geq 0\}$ arising in the process are absorbed. Determine the estimator

$$\widehat{J}_\beta(h) = \sum_{k=0}^{\infty} \sum_{i=1}^{\eta_k} h\left(\xi_k^{(i)}\right) / \beta\left(\xi_k^{(i)};\ k\right).$$

Introduce the notations

$$\mathcal{D}_p\varphi(x) = \sum_{y\in\mathcal{X}} Q(x,y)\,\varphi^2(y) - \left(\sum_{y\in\mathcal{X}} Q(x,y)\,\varphi(y)\right)^2, \quad x \in \mathcal{X},$$

$$\mathcal{D}_p\varphi(0) = \sum_{y\in\mathcal{X}} \nu(y)\,\varphi^2(y) - \left(\sum_{y\in\mathcal{X}} \nu(y)\,\varphi(y)\right)^2,$$

$$r_p(\varphi, x) = \sum_{y\in\mathcal{X}} \gamma(x,y)\, Q(x,y)\, [\varphi(y)\,\beta(x)/\beta(y)]^2, \quad x \in \mathcal{X},$$

$$r_p(\varphi, 0) = \sum_{y\in\mathcal{X}} \gamma(y)\,\nu(y)\,[\varphi(y)/\beta(y)]^2,$$

$$r_p(x,k) = \sum_{y\in\mathcal{X}} \gamma_k(x,y)\, Q(x,y)\, [\varphi(y)\,\beta(x,k)/\beta(y;k-1)]^2,$$

$$r_p(0) = \sum_{y\in\mathcal{X}} \gamma(y)\,\nu(y)\,\varphi^2(y)/\beta^2(y;0),$$

where

$$\gamma(x,y) = \overline{\beta(y)/\beta(x)} - \left(\overline{\beta(y)/\beta(x)}\right)^2,$$
$$\gamma(y) = \overline{1/\beta(y)} - \left(\overline{1/\beta(y)}\right)^2,$$
$$\beta(x), \quad x \in \mathcal{X} \text{ – are positive numbers},$$
$$\widehat{T}_\beta = \sum_{k=0}^{\infty} \eta_k, \quad \widehat{T}_\beta(N) = \sum_{k=0}^{N} \eta_k,$$

Let us use formulas (2.1) and (2.2). Suppose that

$$\beta(y,k) \leq \beta(y), \quad k = 0,1,2,\dots, \tag{3.1}$$

$$\nu^T W \beta < \infty, \quad \nu^T W |h| < \infty. \tag{3.2}$$

The following theorem is valid.

Theorem 3.1. *For any positive integers $\beta(x;k)$, $x \in \mathcal{X}$, $k = 0, 1, \dots$ satisfying (3.1) and (3.2), for an absorbing homogeneous chain with a discrete state space*

1) the quantity $\widehat{J}_\beta$ is an unbiased estimate of the functional $J(h)$. Besides

$$\mathcal{D}\widehat{J}_\beta(h) = \sum_{k=1}^{\infty} \sum_{y \in \mathcal{X}} \left(\nu^T Q^k\right)(y)\, \beta^{-1}(y,k)\, [\mathcal{D}_p \varphi(y) + r_p(y,k)] + \\ + \mathcal{D}_p \varphi(0) + r_p(0),$$

$$E\widehat{T}_\beta = \sum_{k=1}^{\infty} \nu^T Q^k\, \beta(k) + 1;$$

2) if $\beta(y;k) \equiv \beta(y)$, $k = 0, 1, \dots$, $y \in \mathcal{X}$, then

$$\mathcal{D}\widehat{J}_\beta(h) = \sum_{y \in \mathcal{X}} \left(\nu^T W\right)(y)\, \beta^{-1}(y)\, [\mathcal{D}_p \varphi(y) + r_p(y)],$$

$$E\widehat{T}_\beta = \nu^T W \beta + 1;$$

3) if, in addition, $\nu(x) = p(x)$ and an accompanying chain is an ergodic chain, then

$$\mathcal{D}\widehat{J}_\beta(h) = (1/\pi(0)) \sum_{y \in \mathcal{X}} \pi(y)\, \beta^{-1}(y)\, [\mathcal{D}_p \varphi(y) + r_p(\varphi, y)],$$

$$E\widehat{T}_\beta = (1/\pi(0)) \sum_{y \in \mathcal{X}} \pi(y)\, \beta(y).$$

Proof. 1) According to Theorem 2.1,

$$E\widehat{T}_\beta(N) = \sum_{k=1}^{N-1} \nu^T Q^k\, \beta^{(k)} + 1.$$

As the matrix $(I - Q)^{-1}$ exists and $\beta(x;k) \le \beta(x)$, then

$$\sum_{k=1}^{N} \nu^T Q^k\, \beta(k) \le \nu^T \left(\sum_{k=1}^{N} Q^k\right) \beta \le \\ \le \nu^T \left(\sum_{k=1}^{\infty} Q^{k-1}\right) \beta = \nu^T (I - Q)^{-1} \beta < \infty.$$

Therefore

$$E\widehat{T}_\beta = \lim_{N \to \infty} E\widehat{T}_\beta(N) \le \nu^T W \beta < \infty,$$

and hence, the branching process is stopped with probability equal to 1.

In addition,

$$E\widehat{T}_\beta = 1 + \sum_{k=1}^{\infty} \nu^T Q^k \, \beta(k).$$

Further, if $h(x) \geq 0$, then

$$\varphi(x; N-k) = h(x) + \sum_{i=k}^{N} \left(Q^{i-k+1} h\right)(x) =$$

$$= \sum_{i=0}^{N-k} \left(Q^i h\right)(x) \leq \sum_{i=0}^{\infty} \left(Q^i h\right)(x) = \left((I-Q)^{-1} h\right)(x) = \varphi(x),$$

hence

$$\lim_{N\to\infty} \varphi(x; N-k) = \lim_{N\to\infty} \sum_{i=0}^{N-k} \left(Q^i h\right)(x) = \varphi(x). \tag{3.3}$$

Evidently, for alternating $h(x)$ the statement (3.3) is valid under the condition that $\nu^T (I-Q)^{-1}|h| < \infty$ for arbitrary probability measure ν.

Hence

$$\lim_{N\to\infty} \mathcal{D}_p \varphi(x; N-k) = \mathcal{D}_p \varphi(x). \tag{3.4}$$

As

$$\mathcal{D}\widehat{J}_\beta(h) = \lim_{N\to\infty} \mathcal{D}\widehat{J}_\beta(h; N),$$

then substituting equalities (3.3) and (3.4) into the expression for $\mathcal{D}\widehat{J}_\beta(h;N)$ from Theorem 2.1, we get the required statement for $\mathcal{D}\widehat{J}_\beta(h)$. The nonbias of the estimator $\widehat{J}_\beta(h)$ follows from the nonbias of each term of the expansion according to the step number.

2) The corresponding expressions are obtained by direct substitutions $\beta(x)$ instead of $\beta(x;k)$.

3) To make the proof, it is enough to check that

$$p^T (I-Q)^{-1} = \{\pi(y)/\pi(0)\}^T.$$

Designate $\pi' = (\pi(1), \pi(2), \dots)$. From $\pi^T P = \pi^T$, we have

$$\pi(0) p^T = (\pi')^T - (\pi')^T Q,$$
$$(\pi')^T (I-Q) = \pi(0) p^T.$$

Multiplying the latter equality by $(I-Q)^{-1}$, we have

$$(\pi')^T = \pi(0)\, p^T (I-Q)^{-1},$$
$$(\pi'/\pi(0))^T = p^T (I-Q)^{-1}$$

Q.E.D.

4. The steady–state investigation

Let $\{\xi_k;\ k \geq 0\}$ be a regular Markov chain with a discrete state space $\mathcal{X}' = \{0, 1, \dots, n\}$ or $\mathcal{X}' = \{0, 1, \dots\}$, a stationary distribution $\pi = \{\pi(x)\},\ x \in \mathcal{X}'$, and a transition matrix $P = \{P(x, y)\},\ x, y \in \mathcal{X}'$. Let the functional

$$J_{st}(h) = \pi^T h = (\pi, h) = \sum_{x \in \mathcal{X}} \pi(x)\, h(x),$$

be evaluated where $h(x)$ is a given real function.

Construct the estimators of the path branching technique based on the two approaches of the estimation of linear functionals considered in the previous chapter (the regeneration approach and the use of embedded chains).

Regeneration approach.
In the case of a finite state space, it is known that any state is a regenerative state and the returning-time expectation for any state is finite [22]. For a countable state space, we suppose that returning into state $\{0\}$ time expectation is finite and

$$\left((I - Q)^{-1} |h|\right)(x) < \infty, \quad x \in \mathcal{X}, \tag{4.1}$$

where $Q = \{P(x, y)\}_{x, y \in \mathcal{X}}$.

Let $\beta = \{\beta(x)\}_{x, y \in \mathcal{X}}$ be an arbitrary vector with positive coordinates, $\xi_0 = 0$, $\beta(0) = 1$.

Assume that

$$\sum_{x \in \mathcal{X}} \pi(x)\, \beta(x) < \infty. \tag{4.2}$$

In the case of a finite state space, fulfillment of this condition is evident.

Apply the branching procedure described in Sections 2 and 3 to a given chain simulation with $\nu(x) = P(0, x)$, $x \in \mathcal{X}'$, $\beta(x, k) = \beta(x)$. Then any copy simulation is stopped when hitting to $\{0\}$. Define this simulation procedure as a cycle.

Define the estimate $J_\beta(h)$ by the formula

$$\tilde{J}_\beta(h; m) = \left(\sum_{i=1}^{m} Y_\beta(h, j)/m + h(0) \right) \Big/ \left(\sum_{i=1}^{m} Y_\beta(\bar{h}, j)/m + \bar{h}(0) \right), \tag{4.3}$$

where

$$Y_\beta(h, j) = \sum_{k=0}^{\infty} \sum_{i=1}^{\eta_k(j)} h\left(\xi_k^{(i,j)}\right) / \beta\left(\xi_k^{(i,j)}\right),$$

$$Y_\beta(\bar{h}, j) = Y_\beta(h, j) \ \text{ for } \ h = \bar{h}, \quad \bar{h}(x) \equiv 1, \quad x \in \mathcal{X}',$$

$$\eta_k(j) = \eta_k, \quad \xi_k^{(i,j)} = \xi_k^{(i)} \ \text{ in a cycle with number } j.$$

Set

$$\tilde{T}_\beta(m) = \sum_{i=1}^{m} \sum_{k=1}^{\infty} \eta_k(i).$$

Note that $Y_\beta(h,j)$ coincides with the estimate $\widehat{J}_\beta(h)$ for a breaking chain with a matrix $Q = (P(x,y))_{x,y\in\mathcal{X}}$. Hence, according to Theorem 3.1,

$$EY_\beta(h,j) = \sum_{x\in\mathcal{X}} \pi(x)\, h(x)/\pi(0),$$

$$E\sum_{j=1}^{m} Y_\beta(h,j)/m + h(0) = \sum_{x\in\mathcal{X}} \pi(x)\, h(x)/\pi(0) + h(0) = \sum_{x\in\mathcal{X}'} \pi(x)\, h(x)/\pi(0),$$

$$E\sum_{i=1}^{m} Y_\beta(\bar{h},j)/m + \bar{h}(0) = \sum_{x\in\mathcal{X}} \pi(x)/\pi(0) + 1/\pi(0),$$

$$\mathcal{D}Y_\beta(h,j) = \sum_{x\in\mathcal{X}} \pi(x)\,\beta^{-1}(x)\,[\mathcal{D}_p\varphi(x) + r_p(\varphi,x)]\,/\pi(0).$$

Designate

$$\tilde{h} = \{\tilde{h}(x)\}_{x\in\mathcal{X}'},$$
$$\tilde{h}(x) = h(x) - J_{st}\bar{h}(x),$$
$$\tilde{\varphi}(x) = (I-Q)^{-1}\tilde{h}.$$

From Theorem 3.1, we have

$$\mathcal{D}Y_\beta(\tilde{h},j) = \sum_{x\in\mathcal{X}'} \pi(x)\,\beta^{-1}(x)\,[\mathcal{D}_p\tilde{\varphi}(x) + r_p(\tilde{\varphi},x)]\,/\pi(0). \tag{4.4}$$

The following lemma is valid.

Lemma 4.1. *For $m\to\infty$, the distribution of the random variable $\sqrt{m}\,\tilde{J}_\beta(h,m)$ converges to normal distribution with parameters $(J_{st},\ \sigma_0^2)$ where $\sigma_0^2 = \mathcal{D}Y_\beta(\tilde{h};1)\,\pi^2(0)$.*

Proof. The quantity $\tilde{J}_\beta(h,m)$ is represented in the form

$$\bar{\theta} = \left(\sum_{i=1}^{m} Z_i/m\right) \Big/ \left(\sum_{i=1}^{m} v_i/m\right),$$

where $\{Z_i\}$, $\{v_i\}$ are independent, equally distributed random variables. According to the Central Limit Theorem, the $\sqrt{m}\,\bar{\theta}$ distribution converges to normal distribution with parameters $\theta = E\bar{\theta}$,

$$(Ev_1)^{-2}\,\mathcal{D}(Z_1) = \mathcal{D}Y_\beta(\tilde{h};1)/\left(EY_\beta(\bar{h};1)\right)^2 = \mathcal{D}Y_\beta(\tilde{h};1)\,\pi^2(0).$$

The next theorem follows from Lemma 4.1 and expression (4.4) for $\mathcal{D}Y_\beta(\tilde{h};1)$.

Theorem 4.1. *For a regular Markov chain with a finite state space and for an ergodic chain with a countable state space and with a finite return into* $\{0\}$ *time expectation, if this chain satisfies condition (4.1),* $\tilde{J}_\beta(h,m)$ *is an asymptotically unbiased estimator* J_{st} *at any positive values of* $\beta(x)$, $x \in \mathcal{X}$ *satisfying condition (4.2). Besides*

$$\lim_{m\to\infty} m^{-1}\mathcal{D}\tilde{J}_\beta(h,m) = \pi(0)\sum_{x\in\mathcal{X}} \pi(x)\,\beta^{-1}(x)\,[\mathcal{D}_p(\tilde{\varphi}(x)) + r_p(\tilde{\varphi},x)],$$

$$E\tilde{T}_\beta(m) = m\sum_{x\in\mathcal{X}} \pi(x)\,\beta(x)/\pi(0).$$

Note that the quantity $\mathcal{D}J_\beta ET_\beta$, where $\mathcal{D}J_\beta(h) = \lim_{m\to\infty} m^{-1}\mathcal{D}\tilde{J}_\beta(h,m)$, $ET_\beta = E\tilde{T}_\beta(1)$, has the form

$$\mathcal{D}J_\beta(h)ET_\beta = \sum_{x\in\mathcal{X}'} \pi(x)\,\beta(x)\sum_{x\in\mathcal{X}} \pi(x)\,\beta^{-1}(x)\,[\mathcal{D}_p\tilde{\varphi}(x) + r_p(\tilde{\varphi},x)] \tag{4.5}$$

by Theorem 4.1.

An embedded chain application.
Let a sequence of random processes

$$\left\{\xi_k^{(i)}(j),\quad i\in 1:\eta_k(j),\quad k=0,1,\dots,N\right\},\quad j=1,2,\dots,m$$

be simulated by the algorithm described in Section 2, with

$$\xi_0(1) = x \ \text{ with probability }\ \nu(x),\quad x\in\mathcal{X}',$$

$$\xi_0(j) = \xi_N^{(i)}(j-1)\ \text{ with probability }\ \beta\left(\xi_N^{(i)}(j-1);\,N\right) \Big/ \sum_{\gamma=1}^{\nu_N(j-1)} \beta\left(\xi_N^{(\gamma)}(j-1);\,N\right),$$

where $\beta(x;k)$, $x\in\mathcal{X}'$, $k=0,1,\dots,N$ are given nonnegative numbers, $\mathcal{X}$ is replaced by $\mathcal{X}'$, and formulas (2.1) – (2.2) are used, and η_k is replaced by $\eta_k(j)$, $\xi_k^{(i)}$ is replaced by $\xi_k^{(i)}(j)$. Determine the estimator

$$\widehat{J}_\beta(h;\,N,m) = \frac{1}{m}\sum_{j=1}^{m}\sum_{k=0}^{N}\sum_{j=1}^{\eta_k(i)} h\left(\xi_k^{(\gamma)}\right)\,/\,\beta\left(\xi_k^{(\gamma)};\,k\right).$$

Let

$$\widehat{T}_\beta(N;m) = \sum_{j=1}^{m}\sum_{k=0}^{N}\eta_k(j).$$

The following result arise directly from Theorem 3.1.

Theorem 4.2. *Let $\{\xi_k;\ k \geq 0\}$ be a regular Markov chain with a discrete state space $\mathcal{X}' = \{0, 1, \dots\}$. For any positive numbers $\beta(x;k)$, $x \in \mathcal{X}'$, $k = 0, 1, \dots, N$ at $m \to \infty$, the quantity $\widehat{J}_\beta(h;\ N, m)$ is asymptotically unbiased and an asymptotically normal estimator of the functional J_{st}. Here*

$$\sqrt{m}\,\mathcal{D}\widehat{J}_\beta(h;\ N,m) \Big/ \sum_{j=1}^{m}\sum_{k=1}^{N}\sum_{y\in\mathcal{X}'} \left(\nu^T P^{(jN+k-1)}\right)(y)\,\beta^{-1}(y,k)\times$$
$$\times\left[\mathcal{D}_p\varphi(y;N-k) + r_p(y;k)\right] \to 1,$$
$$E\widehat{T}_\beta(N;m) = \sum_{j=1}^{m}\sum_{k=1}^{N}\left(\nu^T P^{(jN+k-1)}\right)(y)\,\beta(y,k).$$

5. The optimal design concept

To construct an optimum criterion, we will suppose that branching expenditures can be ignored, and we will consider that the laboriousness of the calculations is determined by the total number of steps of all simulated copies. Thus, estimating one of the linear functionals $J(h,N)$, $J(h)$ or $J_{st}(h)$, we can take the quantity

$$\mathcal{D}\widehat{J}_\beta E\widehat{T}_\beta,$$

where $\widehat{J}_\beta = \widehat{J}_\beta(h;N)$, $\widehat{J}_\beta(h)$, $\tilde{J}_\beta(h;m)$, resp. as a criterion.

Progress in the investigation of optimal designs for classical regression analysis has been achieved to the great extent due mainly to the introduction of so-called approximate designs, which are achieved by the replacement of a discrete measure with a general one. An analogous step is worthwhile considering for designing simulation experiments with branching paths. Ignoring the $r_p(x;k)$ terms in the expression for $\mathcal{D}\widehat{J}_\beta(h;N)$ and the corresponding terms in the representations for $\mathcal{D}\widehat{J}_\beta(h)$ and $\mathcal{D}\tilde{J}_\beta(h;m)$, a convenient representation for an optimal design can be obtained. Those terms depend on β in a complicated way and do not permit the development of a universal and computationally comfortable approach (in fact, in the cases of waiting and of a random walk, we can investigate the asymptotic behavior of these terms and construct designs close to the optimal designs (see Section 10)). By virtue of Note 2.1 above, when using formula (2.3), those terms have order $O(1/L)$ and with L increasing, become relatively small. Note that using formula (2.3) imposes supplementary restrictions on the simulation algorithm and the dependence between copies appears when calculating the number of descendants. This dependence implies that for all copies, a recurrent step calculation is carried out in parallel (or in quasi-parallel).

Ignoring the r_p terms can be grounded in another way. Using formula (2.2), as is shown by Note 2.2, the quantity $r_p(x;k)$ has a simple form, and for the expression of $\mathcal{D}\widehat{J}_\beta ET_\beta$ (2.11), derived from the expression to be investigated further by replacing $\mathcal{D}_p\varphi(x;k)$ or

$\mathcal{D}_p\tilde{\varphi}(x)$ by an expression close to it and independent of β, takes place. Hence, further results are automatically extended to this case. The quantity r_p can be reduced to zero when using the labelling technique for the special type chain investigated in Section 11 (efficiency waste connected with this term is investigated there too).

By virtue of these remarks, let us consider the quantity

$$\sum_{x\in\mathcal{X}'} \pi(x)\,\beta(x) \sum_{x\in\mathcal{X}'} \pi(x)\,\beta^{-1}(x)\,\mathcal{D}_p\tilde{\varphi}(x)$$

as the optimum criterion in the cases of a stationary behaviour and analogous representations for a transition behaviour and absorbing chains. As the stationary behaviour appears to be more important, and its expression is more compact, we will precise provide reasoning for these cases. Analogous results for the transition behaviour and for absorbing chains can be obtained by the same reasoning.

Let $\{\xi_k;\ k \geq 0\}$ be a regular Markov chain with a discrete state space $\mathcal{X}'$ $(\mathcal{X}' = \{0,1,\dots\}$ or $\mathcal{X}' = \{0,1,\dots,n\}$. If a state space is countable, we will assume that the return into state $\{0\}$ time expectation is finite and condition (4.1) is fulfilled.

Let us consider that $\beta(0) = 1$ and introduce the following definition.

Definition 5.1. A vector of positive numbers $\beta = (\beta(0),\beta(1),\dots)$ satisfying condition (4.1), and condition $\beta(0) = 1$ will be called a Branching Simulation Design (BSD).

Optimum $R(h)$ criterion.
Define the quantity $R_\beta(h)$ by the formula

$$R_\beta(h) = \sum_{x\in\mathcal{X}'} \pi(x)\,\beta(x) \sum_{x\in\mathcal{X}'} \pi(x)\,\beta^{-1}(x)\,\mathcal{D}_p\tilde{\varphi}(x).$$

Definition 5.2. A branching simulation design β^* will be called a $R(h)$-optimal if

$$R_{\beta^*}(h) = \inf R_\beta(h),$$

where the lower bound is taken over the set of all branching simulation designs.

The following theorem is valid.

Theorem 5.1. *For an arbitrary regular Markov chain with a discrete state space and a finite of return into $\{0\}$ time expectation and a function $h(x)$, if this chain satisfies condition (4.1), the function $R_\beta(h)$ is strictly convex and a unique $R(h)$-optimal BSD exists. It has the form*

$$\beta^*(x) = \left(\mathcal{D}_p\tilde{\varphi}(x)\,/\,\mathcal{D}_\nu\tilde{\varphi}(0)\right)^{1/2}.$$

Proof. The strict convexity of the function $R_\beta(h)$ follows from the inequality

$$(\alpha a + (1-\alpha)b)^{-1} \leq \alpha a^{-1} + (1-\alpha)b^{-1},$$

being valid for any positive numbers a, b and $0 \le \alpha \le 1$. Existence, uniqueness and an explicit form of the optimal design are derived from the Cauchy–Schwartz inequality in the form

$$\sum_{x \in \mathcal{X}'} a(x)\,\beta(x) \sum_{x \in \mathcal{X}'} b(x)\,\beta^{-1}(x) \le \left(\sum_{x \in \mathcal{X}'} \sqrt{a(x) b(x)} \right)^2 .$$

on the assumption of $a(x) = \pi(x)$, $b(x) = \pi(x)\,\mathcal{D}_p\tilde{\varphi}(x)$, $x \in \mathcal{X}'$.

The proof is complete.

Note that for β^* from, Theorem 5.1 an asymptotically unbiased estimate based on simulation results can be determined.

Let us consider that $a/b = 0$, if $a = 0$ and $b = 0$. Designate

$$\widehat{\mathcal{D}}_p\tilde{\varphi}(x) = \sum_{\gamma=1}^{m} \left[\left(\varphi_1^{(\gamma)}\right)^2 - Y^2(h,\gamma) \right] \chi\left(\xi_0^{(1,\gamma)} = x\right) \Big/ \sum_{\gamma=1}^{m} \chi\left(\xi_0^{(1,\gamma)} = x\right),$$

where

$$\varphi_1^{(\gamma)} = \sum_{k=0}^{\infty} \sum_{i=1}^{\eta_k(\gamma)} h\left(\xi_k^{(i,\gamma)}\right) \big/ \beta\left(\xi_k^{(i,\gamma)}\right), \quad x \ne 0,$$

$$\widehat{\mathcal{D}}_p\tilde{\varphi}(0) = \sum_{\gamma=1}^{m} \chi\left(\xi_0^{(1,\gamma)} = x\right) Y_\beta^2(h,\gamma) - \left(\sum_{\gamma=1}^{m} Y_\beta(h,\gamma)\,\chi\left(\xi_0^{(1,\gamma)} = x\right) \right)^2 .$$

Theorem 5.2. *Under the conditions of Theorem 5.1,*

$$E\widehat{\mathcal{D}}_p\tilde{\varphi}(x) = \mathcal{D}_p\tilde{\varphi}(x)$$

and, with probability 1,

$$\lim_{m\to\infty} \widehat{\mathcal{D}}_p\tilde{\varphi}(x) \,/\, \widehat{\mathcal{D}}_p\tilde{\varphi}(0) = \beta^*(x), \quad x \in \mathcal{X},$$

where β^* is an $R(h)$-optimum branching simulation design.

Proof. The first statement of the theorem is demonstrated by the immediate calculation by Wald's Lemma. The second statement follows from the Strong Law of Large Numbers.

The next result shows that, in the case of a finite state space, quantity $R_\beta(h)$ does not depend on the choice of regeneration state.

Let $\{\xi_k;\ k \ge 0\}$ be a finite regular Markov chain with a state space $\mathcal{X} = \{0, 1, \dots, n\}$. Designate $Q_{(l)} = (P(x,y))_{x,y \ne l}$, $l = 0, 1, \dots, n$. In this case, the return-time expectation is finite for any state $\{l\}$ and the matrix $I - Q_{(l)}$ is invertible.

Suppose $\beta(l)=1$. We can determine estimator $\tilde{J}_\beta(h,m)$ with the regeneration state $\{l\}$ by analogy with the construction in Section 4 and obtain an analogue of Theorem 4.1 by replacing Q with $Q_{(l)}$:

$$\lim_{m\to\infty} m^{-1}\mathcal{D}\tilde{J}_\beta(h,m)=$$
$$=\sum_{x\in\mathcal{X}'}\pi(x)\,\beta(x)\sum_{x\in\mathcal{X}'}\pi(x)\,\beta^{-1}(x)\left[\mathcal{D}_p\tilde{\varphi}_{(l)}(x)+r_p(\tilde{\varphi}_{(l)},x)\right].$$

where

$$\tilde{\varphi}_{(l)}=(I-Q_{(l)})^{-1}\tilde{h}_{(l)},\quad \tilde{h}_{(l)}=(\tilde{h}(0),\dots,\tilde{h}(l-1),\,\tilde{h}(l+1),\dots,\tilde{h}(n)),$$

$$\mathcal{D}_p\tilde{\varphi}_{(l)}(x)=\sum_{y\in\mathcal{X}'\setminus\{l\}}P(x,y)\,\tilde{\varphi}^2_{(l)}(y)-\left(\sum_{y\in\mathcal{X}'\{l\}}P(x,y)\,\tilde{\varphi}_{(l)}(y)\right)^2,$$

$$r_p(\tilde{\varphi}_{(l)},x)=\sum_{y\in\mathcal{X}'\{l\}}\gamma(x,y)\,P(x,y)\,\left[\tilde{\varphi}_{(l)}(y)\,\beta(x)/\beta(y)\right]^2.$$

Rejecting the terms $r_p(\tilde{\varphi}_{(l)},x)$, we have

$$R_\beta(h,l)=\sum_{x\in\mathcal{X}'}\pi(x)\,\beta^{-1}(x)\,\mathcal{D}_p\tilde{\varphi}_{(l)}(x). \tag{5.1}$$

Theorem 5.3. *For a regular Markov chain with a finite state space $\mathcal{X}'=\{0,1,\dots,n\}$, the value $R_\beta(h,l)$ coincides with $R_\beta(h,0)=R_\beta(h)$ for any $l\in\mathcal{X}'$.*

Proof. Represent the quantity

$$\sum_{x\in\mathcal{X}'}\pi(x)\,\beta^{-1}(x)\,\mathcal{D}_p\tilde{\varphi}_{(l)}(x)$$

in the form

$$h^T_{(l)}\left(I-Q^T_{(l)}\right)^{-1}\left(\sum_{x\in\mathcal{X}'}\pi(x)\,\beta^{-1}(x)\,(P(x,y)\sigma_{yz}-P(x,y)\,P(x,z))\right)_{y,z\in\mathcal{X}'\setminus\{l\}}\times$$
$$\times\left(I-Q_{(l)}\right)^{-1}h_{(l)}=$$
$$=\sum_{x\in\mathcal{X}'}\pi(x)\,\beta^{-1}(x)\,h^T_{(l)}\left(I-Q^T_{(l)}\right)^{-1}(P(x,y)\sigma_{yz}-P(x,y)\,P(x,z))\,\left(I-Q_{(l)}\right)^{-1}h_{(l)}.$$

Hold x fixed and designate $a_y = P(x,y),\ y = 0,1,\dots,n$,

$$A_{(l)} = \operatorname{diag}(a_0,\dots,a_{l-1},a_{l+1},\dots,a_n),$$
$$A = \operatorname{diag}(a_0,\dots,a_n),$$
$$a_{(l)} = (a_0,\dots,a_{l-1},a_{l+1},\dots,a_n)^T,\quad a = (a_0,\dots,a_n)^T,$$
$$E = E_l = (l_{x,y})_{x,y\in\mathcal{X}'\setminus\{l\}},\quad l_{x,y}\equiv 1,\quad v = (\delta_{x,y} - P(x,y))_{y\in\mathcal{X}\setminus\{l\}},$$
$$(I-P)_{(l)} = (\delta_{ij} - P(i,j))_{i\in\mathcal{X}'\setminus\{l\},\, j\in\mathcal{X}'}.$$

Let $P(x,y) \neq 0$ for $y \in \mathcal{X}'$. Then the following lemma is valid.

Lemma 5.1. *The equality*

$$(I-P)_{(l)}\,A^{-1}(I-P)_{(l)}^T = \left(I - Q_{(l)}\right)\left(A_{(l)} - a_{(l)}a_{(l)}^T\right)^{-1}\left(I - Q_{(l)}\right)^T$$

is valid for the matrices introduced above.

Proof. The expression

$$(I+BC)^{-1} = I - B(I+CB)^{-1}C$$

is valid for any matrices B and C of a related order (see e.g., [38]). Assume $B = a_{(l)}$, $C = a_{(l)}^T A_{(l)}^{-1}$. Using the aforesaid formula, we have

$$\left(A_{(l)} - a_{(l)}a_{(l)}^T\right)^{-1} = A_{(l)}^{-1}\left(I - a_{(l)}a_{(l)}^T A_{(l)}^{-1}\right)^{-1} =$$
$$= A_{(l)}^{-1} + A_{(l)}^{-1}a_{(l)}a_{(l)}^T A_{(l)}^{-1}\Big/\left(1 - a_{(l)}^T A_{(l)}a_{(l)}\right).$$

Moreover,

$$a_{(l)}^T A_{(l)}^{-1} a_{(l)} = \sum_{i\neq l} a_i = \sum_{y\neq l} P(x,y) = 1 - P(x,l) = 1 - a_l,$$
$$A_{(l)}^{-1}a_{(l)}a_{(l)}^T A_{(l)}^{-1} = E_l.$$

Hence

$$\left(A_{(l)} - a_{(l)}a_{(l)}^T\right)^{-1} = A_{(l)}^{-1} + E/a_l.$$

Further,

$$\left(I - Q_{(l)}\right)E\left(I - Q_{(l)}\right)^T =$$
$$= \left(\sum_{j\in\mathcal{X}'\setminus\{l\}} (\delta_{ij} - P(i,j))\right)\sum_{k\in\mathcal{X}'\setminus\{l\}} (\delta_{sk} - P(s,k))_{i,s\in\mathcal{X}'\setminus\{l\}} = vv^T,$$
$$(I-P)_{(l)}\,A^{-1}(I-P)_{(l)}^T = \left(I - Q_{(l)}\right)A_{(l)}^{-1}\left(I - Q_{(l)}\right)^T + vv^T/a_l,$$
$$\left(I - Q_{(l)}\right)\left(A_{(l)} - a_{(l)}a_{(l)}^T\right)^{-1}\left(I - Q_{(l)}\right)^T =$$
$$= \left(I - Q_{(l)}\right)\left(A_{(l)}^{-1} + E/a_l\right)\left(I - Q_{(l)}\right)^T = \left(I - Q_{(l)}\right)A_{(l)}^{-1}\left(I - Q_{(l)}\right)^T + vv^T/a_l.$$

Thus,

$$(I-P)_{(l)}\, A^{-1}(I-P)_{(l)}^T = \left(I-Q_{(l)}\right)\left(A_{(l)} - a_{(l)}a_{(l)}^T\right)^{-1}\left(I-Q_{(l)}\right)^T.$$

Q.E.D.

The invertibility of matrix $(I-P)_{(l)}\, A^{-1}(I-P)_{(l)}^T$ (for $P(x,y) \neq 0,\ \ y \in \mathcal{X}'$) follows from Lemma 5.1. Indeed,

$$(I-P)_{(l)}\, A^{-1}(I-P)_{(l)}^T = \left(I-Q_{(l)}\right)\, A_{(l)}^{-1}\left(I-Q_{(l)}\right)^T + vv^T/a_l.$$

The invertibility of the first matrix follows from the invertibility of $\left(I-Q_{(l)}\right)$. As for the second one, it is nonnegatively definite, hence the determinant of the matrix

$$(I-P)_{(l)}\, A^{-1}(I-P)_{(l)}^T$$

is not equal to zero.

Lemma 5.2. *For the introduced matrices, the quantity*

$$R_{(l)} = \pi(l)\, \tilde{h}_{(l)}^T \left[(I-P)_{(l)}\, A^{-1}(I-P)_{(l)}^T\right]^{-1} \tilde{h}_{(l)}$$

does not depend on l.

Proof. Produce $R_{(l)}$ in the form

$$R_{(l)} = \pi(l)\, \det\begin{pmatrix} 1 & \tilde{h}_0 \ldots \tilde{h}_{l-1}\, \tilde{h}_{l+1} \ldots \tilde{h}_n \\ \tilde{h}_{(l)} & (I-P)_{(l)}\, A^{-1}(I-P)_{(l)}^T \end{pmatrix} \Big/ \det\left[(I-P)_{(l)}\, A^{-1}(I-P)_{(l)}^T\right] - \pi(l),$$

where $\tilde{h}_i = \tilde{h}(i)$. Adding the columns corresponding to $\tilde{h}_i$ and multiplied by π_i, to the second column, then adding the column corresponding to the h_i lines multiplied by π_i $(i = 0, 1, \ldots, l-1, l+1, \ldots, n)$ to the second line and changing over the second and first columns and the second and first lines in the obtained determinant, we have

$$R_{(l)} = R_{(0)},$$

as

$$\sum_{i \neq l} \tilde{h}_i\, \pi(i) = h(l)\, \pi(l),$$

$$\sum_{j \neq l} (\delta_{ik} - P(i,k))\, a_k^{-1}(\delta_{jk} - P(j,k))\, \pi(j) =$$

$$= a_k^{-1}(\delta_{ik} - P(i,k))\,(\delta_{lk} - P(l,k))\, \pi(l).$$

Q.E.D.

Let us provisionally ignore the convention $\beta(l) = 1$. Under the condition $P(x,y) \neq 0$, $x, y \in \mathcal{X}'$ from Lemma 5.2 and representation (5.1), it follows that the quantity $R_\beta(h,l)$ is independent of l. If some $P(x,y)$ are equal to zero, the $R_\beta(h,l)$ independence of l is demonstrated by the passage to the limit. The quantity $R_\beta(h,l)$ does not change when multiplying the vector β by an arbitrary scalar. Hence, the convention $\beta(l) = 1$ does not alter the value of $R_\beta(h,l)$. The design $\beta' = \beta/\beta(0)$ for regeneration state $\{0\}$ corresponds to the design β for regeneration state $\{l\}$. Q.E.D.

The derived result shows that the choice of regeneration state does not affect the choice of design and we can arbitrarily choose it for the convenience of calculation. Let us consider $\{0\}$ as a regeneration state.

6. L-OPTIMUM CRITERION

In regressive experimental design theory, convex functionals (in particular, linear functionals) of a variance-covariance matrix are considered as the optimum criterion. It is advisable to use this approach in simulation experiments for the design of branching paths. Here and in the next Section, we will consider an analogue of the two most-used criteria, namely, the linear criterion and the D-criterion.

Let $\{\xi_k;\ k \geq 0\}$ be a regular Markov chain with a finite state space $\mathcal{X}' = \{0, 1, \ldots, n\}$. Denote by $D_{\tilde{\pi}}$ the matrix

$$D_{\tilde{\pi}} = \lim_{m \to \infty} \left(m^{-1} \operatorname{Cov}(\tilde{\pi}(x), \tilde{\pi}(y)) \right)_{x,y \in \mathcal{X}},$$

where $\tilde{\pi}(x) = \tilde{J}_\beta(h_x; m),\ \ h_x = (\delta_{xy})_{y \in \mathcal{X}}$.

Set

$$r(x,y,z) = \gamma(x,y)\,(\beta(x)/\beta(y))\,(\beta(x)/\beta(z)),$$
$$W = (I - Q)^{-1}, \quad Q = (P(x,y))_{x,y \in \mathcal{X}}.$$

Lemma 6.1. *The matrix $D_{\tilde{\pi}}$ has the form*

$$\mathcal{D}_{\tilde{\pi}} = \mathcal{D} = W^T \left(\sum_{x \in \mathcal{X}'} \pi(x)\,\beta^{-1}(x) \right) [P(x,y)\delta_{xy} - P(x,y)\,P(x,z) + r(x,y,z)]W$$

$(y, z \in \mathcal{X})$ for a finite regular Markov chain.

Proof. From Theorem 4.1, it follows that

$$\lim_{m \to \infty} m^{-1} \mathcal{D}\tilde{J}_\beta(h,m) = \tilde{h}^T \mathcal{D} \tilde{h}.$$

On the other hand,

$$\lim_{m \to \infty} m^{-1} \mathcal{D}\tilde{J}_\beta(h,m) = \tilde{h}^T \mathcal{D}_{\tilde{\pi}} \tilde{h}.$$

Hence

$$\tilde{h}^T \mathcal{D} \tilde{h} = \tilde{h}^T \mathcal{D}_{\tilde{\pi}} \tilde{h}$$

for any vector h.

Let $z = (z(x))_{x\in\mathcal{X}}$ be an arbitrary real vector. Set $h(x) = z(x) - \sum_{x\in\mathcal{X}} z(x)\,\pi(x)/\pi(0)$. Then $\tilde{h}(x) = z(x),\ \ x \in \mathcal{X}$. Therefore,

$$z^T \mathcal{D} z = z^T \mathcal{D}_{\tilde{\pi}} z$$

for any real vector z. As matrices $\mathcal{D}$ and $\mathcal{D}_{\tilde{\pi}}$ are symmetric, then the matrix $\mathcal{D} - \mathcal{D}_{\tilde{\pi}}$ is symmetric too. Therefore, all eigenvalues are real and, since

$$z^T(\mathcal{D} - \mathcal{D}_{\tilde{\pi}})z \equiv 0,$$

are equal to zero. Hence, $\mathcal{D}_{\tilde{\pi}} = \mathcal{D}$. Q.E.D.

Reject the term $r(x, y, z)$ for the reasons pointed out in Section 5 and consider matrices of the form

$$W^T \left(\sum_{x\in\mathcal{X}'} \pi(x)\,\beta^{-1}(x)\,(P(x,y)\delta_{zz} - P(x,y)\,P(x,z)) \right)_{y,z\in\mathcal{X}'} W.$$

As we have to take into account the mean time of simulation $ET_\beta = \sum \beta(x)\,\pi(x)$, let us introduce the quantity

$$\tau(x) = \beta(x)\,\pi(x) \Big/ \sum_{x\in\mathcal{X}'} \pi(x)\,\beta(x)\,.$$

Then $\sum_{x\in\mathcal{X}'} \tau(x) = 1,\ \ \tau(x) > 0$, i.e., $\tau(x)$ is a positive probability measure on $\mathcal{X}'$.

Approximate the variance-covariance matrix, multiplied by the mean simulation time, by the following matrix

$$D(\tau) = \left(I - Q^T\right)^{-1} B(\tau)\,(I - Q)^{-1},$$

where $B(\tau) = \sum_{x\in\mathcal{X}'} \pi^2(x)\,\tau^{-1}\,B_x$,

$$B_x = (P(x,y)\delta_{yz} - P(x,y)\,P(x,z))_{y,z\in\mathcal{X}}\,.$$

Introduce the following definition.

Definition 6.1. A positive probability measure $\tau(x)$ defined on $\mathcal{X}'$ is called an experimental design.

Note that a branching simulation design β can be regenerated by τ in an evident way:

$$\beta(x) = \tau(x)\,\pi(0)\,/\,(\pi(x)\,\tau(0)).$$

Since $\tau(x)$ and $\beta(x)$ determine each other in one-to-one correspondence, let us consider further the problem of the optimal choice of experimental design τ.

Definition 6.2. An experimental design τ^* is called L-optimal if

$$\text{tr } LD(\tau^*) = \inf \text{ tr } LD(\tau),$$

where L is a given nonnegatively definite matrix, and the infimum is taken over the set of all experimental designs.

L-optimal designs may be found in explicit form, for which the following theorem is valid.

Theorem 6.1. *For any finite regular Markov chain the function* tr $LD(\tau)$ *is strictly convex, and an L-optimal design exists. This design is unique and has the form*

$$\tau^*(x) = \pi(x)\sqrt{\text{tr } B_x G} \Bigg/ \sum_{x \in \mathcal{X}'} \pi(y)\sqrt{\text{tr } B_x G}\,,$$

where

$$G = WLW^T.$$

Proof. Note that

$$\text{tr } LD(\tau) = \text{tr } LW^T B(\tau) W = \text{tr } B(\tau) WLW^T = \text{tr } B(\tau) G.$$

Further, as for any real numbers a and b,

$$\frac{1}{(a+b)/2} \leq \frac{1}{2a} + \frac{1}{2b} \tag{6.1}$$

and the equality is attained only in the case of $a = b$, then

$$\sum_x \frac{\pi^2(x)}{(\tau_1(x) + \tau_2(x))/2} \text{ tr } B_x G \leq$$
$$\leq \frac{1}{2} \sum_x \frac{\pi^2(x)}{\tau_1(x)} \text{ tr } B_x G + \frac{1}{2} \sum_x \frac{\pi^2(x)}{\tau_2(x)} \text{ tr } B_x G.$$

Here, equality is attained if and only if $\tau_1(x) = \tau_2(x)$. Since the left side is

$$\text{tr } LD(\tau_1 + \tau_2)/2),$$

and the right side is equal to

$$1/2 \text{ tr } LD(\tau_1) + 1/2 \text{ tr } LD(\tau_2),$$

it follows that tr $LD(\tau)$ is a strictly convex function τ. The latter statement of Theorem 6.1 is derived from the inequality

$$\operatorname{tr} LD(\tau) = \sum_x \pi^2(x)\tau^{-1}(x) \operatorname{tr} B_x G \sum_x \tau(x) \le \left(\sum_x \pi(x)\sqrt{\operatorname{tr} B_x G}\right).$$

Here equality is attained if and only if $\tau(x) = \pi(x)\sqrt{\operatorname{tr} B_x G}$ const. This inequality is a special case of the Cauchy-Schwartz inequality.

The proof is complete.

Consider the problem of estimation of linear functionals as an example of applying the L-criterion.

Let the functionals $J_{st}(h^{(1)}), \dots, J_{st}(h^{(k)})$ are to be estimated. Let us take the quantity $\lim_{m\to\infty} m^{-1}\sum_{l=1}^k c_l D\tilde{J}_\beta(h^{(l)}; m)$ as a measure of estimator accuracy where $c_1, \dots, c_k$ are given coefficients. We have

$$\lim_{m\to\infty} m^{-1}\sum_{l=1}^k c_l D\tilde{J}_\beta(h^{(l)}; m) = \sum_{l=1}^k c_l \left(\tilde{h}^{(l)}\right)^T D_{\tilde{\pi}}\tilde{h}^{(l)} = \operatorname{tr} LD_{\tilde{\pi}},$$

where $L = \sum_{l=1}^k c_l \tilde{h}^{(l)}\left(\tilde{h}^{(l)}\right)^T$.

Replacing the matrix $D_{\tilde{\pi}}$ by the matrix $D(\tau)$ we arrive at the L-criterion.

7. D-OPTIMUM CRITERION

It is well known [31] that the square root of the determinant of a variance-covariance matrix is proportional to the volume concentration ellipsoide of the estimators. Therefore, the quantity

$$\det D(\tau) = \det{}^2 W \det B(\tau).$$

is a natural criterion for optimality of a design τ in the evaluation problem for a stationary distribution of a finite regular Markov chain. From this equality, it follows that the quantity $\det D(\tau)$ infimum over the set of all experimental designs coincides with the infimum of the quantity $\det B(\tau)$ multiplied by $\det^2 W$. Thus, let us introduce the following definition.

Definition 7.1. An experimental design τ^* is called D-optimum if

$$\det B(\tau^*) = \inf \det B(\tau),$$

where the infimum is taken over the set of all experimental designs.

It is impossible to find D-optimum designs in an explicit form but a recurrence procedure of finding them can be constructed. The offered procedure is based on the following Kiefer-Wolfowitz equivalence theorem (see [31, 58]) analogue.

Theorem 7.1. *For any finite regular Markov chain a D-optimum design exists which is unique and satisfies the relation*

$$\tau^*(x) = \left(\text{tr } B_x B^{-1}(\tau^*)\right)^{1/2} \pi(x) \, / \, \sqrt{m}.$$

Moreover, the following three statements are equivalent

$$a) \quad \max_x \frac{\pi^2(x)}{(\tau^*(x))^2} \text{ tr } B_x B^{-1}(\tau^*) = m,$$

$$b) \quad \min_\tau \max_x \frac{\pi^2(x)}{(\tau(x))^2} \text{ tr } B_x B^{-1}(\tau) = \max_x \frac{\pi^2(x)}{(\tau^*(x))^2} \text{ tr } B_x B^{-1}(\tau^*),$$

c) τ^* *– is a D-optimum design.*

Proof. The following lemma is required.

Lemma 7.1. *For any finite regular Markov chain,*

$$\inf \det B(\tau) > 0.$$

Proof of Lemma 7.1. Let $\tau_0 = \pi(x)$, $x \in \mathcal{X}'$. Such a design corresponds to the trivial branching simulation design $\beta_0(x) \equiv 1$. The matrix $D(\tau_0)$ is nonnegatively defined, since it is the estimator's $\tilde{\pi}_i$ variance–covariance matrix for $\beta = \beta_0$. Thus, the matrix $B(\tau_0) = (I - Q)\, D(\tau_0)\, (I - Q)^T$ is nonnegatively defined. Note that the equality $p(x, 0) = 0$, $x = 1, 2, \ldots, n$, is impossible (in this case, the state $\{0\}$ would be irreversible that contradicts the definition of a regular chain). Hence, at least one of the inequalities ($y = 1, 2, \ldots, n$) in

$$\begin{aligned} B(\tau)(y,y) &= \sum_x \pi(x)\, P(x,y) - \sum_x \pi(x)\, P^2(x,y) = \\ &= \sum_x \sum_z \pi(x)\, P(x,y)\, P(x,z) - \sum_x \pi(x)\, P^2(x,y) = \\ &= \sum_x \sum_{z \in \mathcal{X}' \setminus y} \pi(x)\, P(x,y)\, P(x,z) = \\ &= \sum_{z \in \mathcal{X}' \setminus y} \sum_x \pi(x)\, P(x,y)\, P(x,z) \geq \sum_{z \in \mathcal{X}' \setminus y} |(B(\tau_0))(y,z)| \end{aligned}$$

is strict. Thus, Hadamard's weakened conditions are fulfilled and, according to Olga Tuski's theorem [43], the matrix $B(\tau_0)$ is nonsingular. Since the matrix $B(\tau_0)$ is nonnegatively defined, it is positively defined and

$$\det B(\tau_0) > 0.$$

Further, since $\pi(x) > 0,\ \ x \in \mathcal{X}'$ for regular finite chains, then $\min \pi(x) \geq \varepsilon > 0$ for some ε. Moreover, $\tau(x) \leq 1$, since $\tau(x) > 0$ and $\sum \tau(x) = 1$. Therefore,

$$\inf \det B(\tau) = \inf \det \left(\sum_{x \in \mathcal{X}'} \frac{\pi^2(x)}{\tau(x)} B_x \right) \geq$$

$$\geq \det \left(\sum_{x \in \mathcal{X}'} \pi(x) B_x / \varepsilon \right) = \varepsilon^n \det B(\tau_0) > 0.$$

Q.E.D.

From inequality (6.1), it follows that

$$\sum_x \frac{\pi^2(x)}{(\tau_1(x) + \tau_2(x))/2} B_x \leq \frac{1}{2} \sum_x \frac{\pi^2(x)}{\tau_1(x)} B_x + \frac{1}{2} \sum_x \frac{\pi^2(x)}{\tau_2(x)} B_x,$$

i.e.

$$B((\tau_1 + \tau_2)/2) \leq \frac{1}{2} B(\tau_1) + \frac{1}{2} B(\tau_2).$$

Here equality is attained if and only if $\tau_1 = \tau_2$.

Note that by virtue of Lemma 7.1, a set of matrices $B(\tau)$ corresponding to the set of all experimental designs is a subset of positively defined matrices. According to the continuaty of function $\det A$ this subset is bounded and closed under the additional condition $\det B(\tau) \leq C$, where C is a constant.

Let τ^* be a D-optimum design and τ_α be in the form of $\tau_\alpha = (1 - \alpha)\tau^* + \alpha \tau_x$, where $\tau_x(y) = \delta_{xy}$. Since

$$\ln \det B(\tau_\alpha) \geq \ln \det B(\tau^*)$$

for any $\alpha > 0,\ \ x \in \mathcal{X}'$, then

$$\frac{1}{\alpha} [\ln \det B(\tau_\alpha) - \ln \det B(\tau^*)] \geq 0.$$

Turning the quantity α to zero, we have

$$\frac{\partial}{\partial \alpha} \ln \det B(\tau_\alpha) \bigg|_{\alpha = +0} \geq 0.$$

Thus, for any $x \in \mathcal{X}'$

$$\begin{aligned}
&\frac{\partial}{\partial\alpha} \ln\det B(\tau_\alpha)\Big|_{\alpha=+0} = \\
&= \mathrm{tr}\ B^{-1}(\tau^*)\ \frac{\partial B(\tau_\alpha)}{\partial\alpha}\Big|_{\alpha=+0} = \\
&= trB^{-1}(\tau^*) \sum_y \frac{\partial}{\partial\alpha} \left(\tau^*(y)(1-\alpha)+\alpha\tau_x\right)^{-1} \pi^2(y)\, B_y\Big|_{\alpha=+0} = \\
&= trB^{-1}(\tau^*)\, B(\tau^*) - \frac{\pi^2(x)}{(\tau^*(x))^2}\ \mathrm{tr}\ B_x B^{-1}(\tau^*) = \\
&= m - \frac{\pi^2(x)}{(\tau^*(x))^2}\ \mathrm{tr}\ B_x B^{-1}(\tau^*) \geq 0.
\end{aligned}$$

Hence

$$\max_x \frac{\pi^2(x)}{(\pi^*(x))^2}\ \mathrm{tr}\ B_x B^{-1}(\tau^*) \leq m. \tag{7.1}$$

On the other hand,

$$\begin{aligned}
m &= \mathrm{tr}\ B^{-1}(\tau^*)\, B(\tau^*) = \sum_{x\in\mathcal{X}'} \frac{\pi^2(x)}{\tau^*(x)}\ \mathrm{tr}\ B_x B^{-1}(\tau^*) = \\
&= \sum_{x\in\mathcal{X}'} \frac{\pi^2(x)}{(\tau^*(x))^2}\, \tau^*(x)\ \mathrm{tr}\ B_x B^{-1}(\tau^*) \leq \\
&\leq \max_x \frac{\pi^2(x)}{(\tau^*(x))^2}\ \mathrm{tr}\ B_x B^{-1}(\tau^*)
\end{aligned}$$

(since $\sum_{x\in\mathcal{X}'} \tau^*(x) = 1$ and $\sum_{x\in\mathcal{X}'} a(x)\,\tau^*(x) \leq \max_{x\in\mathcal{X}'} a(x)$ for any nonnegative function $a(x)$).

The latter inequality and (7.1) are compatible if and only if

$$\frac{\pi^2(x)}{(\tau^*(x))^2}\ \mathrm{tr}\ B_x B^{-1}(\tau^*) \equiv m, \quad \text{for} \quad x \in \mathcal{X}.$$

Hence, $\tau^*(x) = \pi(x)\left(\mathrm{tr}\ B_x B^{-1}(\tau^*)/m\right)^{1/2}$.

Obviously, D-optimal design exists. To establish that this design is uniquely defined, we will use the inequality (see [38])

$$\left(\frac{A+B}{2}\right)\left(\frac{A^{-1}+B^{-1}}{2}\right) \geq I, \tag{7.2}$$

where $A, B > 0$ and equality takes place if and only if $A = B$. Suppose that there are two D-optimal designs: τ_1 and τ_2.Then, from (7.1), it follows that for any design τ

$$\text{tr } B(\tau)B^{-1}(\tau_1) \leq m, \quad \text{tr } B(\tau)B^{-1}(\tau_2) \leq m$$

and, therefore,

$$\text{tr } B(\tau_2)B^{-1}(\tau_1) \leq m, \quad \text{tr } B(\tau_1)B^{-1}(\tau_2) \leq m,$$

$$\text{tr} \left(\frac{B(\tau_1) + B(\tau_2)}{2}\right)\left(\frac{B^{-1}(\tau_1) + B^{-1}(\tau_2)}{2}\right) \leq m.$$

From the last inequality and (7.2), it follows that $B(\tau_1) = B(\tau_2)$ which is possible if and only if $\tau_1 = \tau_2$.

The equivalence of a), b) and c) is obvious from the above consideration.

The proof of Theorem 7.1 is complete. Q.E.D.

Construct a numerical technique for finding a D-optimal design. Let $\tau_0(y) = \pi(y)$, $y \in \mathcal{X}'$,

$$\tau_{k+1}(y, \alpha) = (1 - \alpha)\tau_k(y) + \alpha\tau_x(y), \quad k = 0, 1, \dots ,$$

where

$$x = \arg\max_x \frac{\pi^2(x)}{\tau_k^2(x)} \text{ tr } B_x B^{-1}(\tau_k).$$

Assume that

$$\alpha_k = \arg \min_{\alpha \in [0,1]} \det B(\{\tau_{k+1}(y, \alpha)\}),$$

$$\tau_{k+1}(y) = \tau_{k+1}(y, \alpha_k), \quad y \in \mathcal{X}'.$$

The following theorem is valid.

Theorem 7.2. *For any finite regular Markov chain,* $\det B(\tau_k) \to \det B(\tau^*)$, *where* τ^* *is a D-optimum design and we can single out a subsequence weakly converging to* τ^* *from the sequence* τ_k.

Proof. Designate $\gamma_k = \ln \det B(\tau_k)$. By its definition, this sequence does not increase. Hence, it has a limit γ^*. By the Helly's Theorem, let us select from τ_k a converging susequence. For the sake of convenience, we designate it in the same manner as the initial one, and its limit – by τ^+. Suppose that $\gamma^* \neq \ln \det B(\tau^*)$. Then $\gamma^* - \ln \det B(\tau^*) > 0$ and $\tau^+ \neq \tau^*$. Since the design τ^* is not a D-optimum design, then by Theorem 7.1, for some $\delta > 0$,

$$\max_x \frac{\pi^2(x)}{(\tau_k(x))^2} \text{ tr } B_x B^{-1}(\tau_k) > m + \frac{\delta}{2}.$$

Expanding the quantity $\ln \det B(\tau_{k+1}(\alpha))$ in the Teylor series with α, we have

$$\gamma_{k+1} - \gamma_k = \alpha_k \frac{\partial}{\partial \alpha} \ln \det B(\tau_k(\alpha)) \,|_{\alpha=+0} + \\ + \left(\alpha_k^2/2\right) \frac{\partial^2}{\partial \alpha^2} \ln \det(\tau_\alpha)\Big|_{\alpha=\alpha^+},$$

where $0 \le \alpha^+ \le 1$.

From the known formulas (see, e.g., [38])

$$\frac{\partial}{\partial \alpha} \ln \det A = \operatorname{tr} \frac{\partial A}{\partial \alpha} A^{-1},$$
$$\frac{\partial A^{-1}}{\partial \alpha} = -A^{-1} \frac{\partial A}{\partial \alpha} A^{-1},$$

it follows that a function $\ln \det A$ is infinitely differentiable on a set of positively defined matrices.

Under the additional condition of $\det B(\tau) \le C$ where C is a constant, as $\det B(\tau) \ge \varepsilon' > 0$ for some ε' (by Lemma 7.1), the set of matrices $\{B(\tau)\}$ is a compact subset of a set of positively defined matrices. It is easy to see that this condition does not confine the process under investigation.

Hence, the function $\ln \det B(\tau_k(\alpha))$ is twice continuously differentiable on a compact set, thus

$$\frac{\partial}{\partial \alpha^2} \ln \det B(\tau_k(\alpha)) \le K < \infty,$$

where K is a positive constant. Now we have

$$\gamma_{k+1} - \gamma_k \le \min_{\alpha \in [0,1]} \left(-\alpha\delta + \frac{\alpha^2}{2} K \right) = -\frac{\delta^2}{2K}$$

for $K \ge N$. Summing up these inequalities for $K = N, N+1, \dots, N+M$, we have

$$\gamma_{N+M} \le \gamma_N - M \frac{\delta^2}{2K},$$

i.e., $\gamma_{N+M} \to -\infty$ for $M \to \infty$. But $\gamma_{N+M} \ge \ln \det B(\tau^+) > 0$. This contradiction shows that $\gamma^* = \ln \det B(\tau^*)$. By virtue of the uniqueness of the optimum design, it follows that $\tau^+ = \tau^*$

Q.E.D.

8. Experimental design for a transient regime and for absorbing chains

Consider the estimator $\widehat{J}_\beta(h; N)$ of the functional $J(h; N)$ defined in Section 2. By virtue of the considerations of Section 5, neglect the terms $r_Q(x; k)$, $r_\nu(0; N)$. From Theorem 2.1, we have the following approximation for $\mathcal{D}\widehat{J}_\beta(h; N)\, E\widehat{T}_\beta(N)$:

$$R_\beta(h; N) = \left[\mathcal{D}_\nu \varphi(0; N) + \sum_{k=1}^{N-1} \sum_{y \in \mathcal{X}} \left(\nu^T Q^{(k)} \right)(y)\, \beta^{-1}(y; k)\, \mathcal{D}_Q\, \varphi(y; N-k) \right] \times$$
$$\times \left[1 + \sum_{k=1}^{N-1} \sum_{y \in \mathcal{X}} \left(\nu^T Q^{(k)} \right)(y)\, \beta(y; k) \right].$$

Introduce the following definitions.

Definition 8.1. A collection of vectors with positive coefficients $\beta(1) = \{\beta(x;1)\}, \ldots,$ $\beta(N) = \{\beta(x;N)\}$, where $x \in \mathcal{X}$, is called a branching simulation design (when evaluating a transient behaviour).

Definition 8.2. A branching simulation design $\beta^* = \{\beta(1), \ldots, \beta(N)\}$ is called an $R(h;N)$-optimum design if

$$R_{\beta^*}(h;N) = \inf R_\beta(h;N),$$

where the infimum is taken over the set of all branching simulation designs.

Let us formulate an analogue of Theorem 5.1.

Theorem 8.1. *For any Markov chain with a discrete state space satisfying condition (2.8), the quantity $R_\beta(h;N)$ is a strictly convex function of vectors $\beta(1), \ldots, \beta(N)$, an $R(h;N)$-optimum branching simulation design exists, is unique, and has the form*

$$\beta^*(y,k) = \sqrt{\mathcal{D}_Q\,\varphi(y;N-k)/\mathcal{D}_\nu\varphi(0,N)}.$$

The proof of Theorem 8.1 is omitted, since it repeats that of Theorem 5.1.

An experiment for designing with absorbing chains.
Consider the estimator $\widehat{J}_\beta(h)$ of the functional $J(h)$ defined in Section 3. Under the conditions of Theorem 3.1 (1), it is shown that

$$\mathcal{D}_Q\,\varphi(y;N-k) \to \mathcal{D}_p\varphi(y).$$

Hence, the elements of the $R(h;N)$-optimum design

$$\beta^*(y,k) \to \sqrt{\mathcal{D}_p\varphi(y)/\mathcal{D}_p\varphi(0)}$$

lose their dependence on k with N increasing.

Therefore, for absorbing chains, it is appropriate to define a branching simulation design using Definition 5.1. The results of Sections 5-7 for a stationary study hold for absorbing chains with apparent alterations.

9. A CHAIN WITH TWO STATES

In applications, a state space is often divided into two subsets (see [86]) that correspond to a Markov chain with a two-states model. In that case, an exhaustive analysis of the efficiency of the branching-path technique is possible. The present Section is devoted to such an analysis.

Let $\{\xi_k;\ k \ge 0\}$ be a regular Markov chain with two states $\{0,1\}$, the transition matrix $P = (p_{ij})_{i,j=0,1}$ and stationary probabilities π_0, $\pi_1 = 1 - \pi_0$. Designate $x = p_{10}$, $y = p_{01}$. Then $p_{11} = 1 - x$, $p_{00} = 1 - y$ and

$$\pi_0(1-y) + (1-\pi_0)x = \pi_0,$$

whence $y = (\pi_1/\pi_0)x$. Thus, the chain is defined by the parameters p_{10} and π_0. Determine the efficiency function by the relation

$$\bar{R}(x, \pi_0) = R_{\beta_0}(h) / \min_{\beta} R_\beta(h)$$

that specifies the optimum design efficiency with respect to the trivial design $\beta_0 = (1,1)^T$. It is evident that the necessary and sufficient chain regularity condition is $0 < x < \min\{1, \pi_0/\pi_1\}$.

Let $0 < x < \min\{1, \pi_0/\pi_1\}$

The following theorem is valid.

Theorem 9.1. *For a regular Markov chain with two states and stationary probabilities π_0, π_1, the efficiency function does not depend on h and has the form*

$$\bar{R}(x, \pi_0) = \left(2 - \frac{x}{\pi_0}\right) \Big/ \left(\sqrt{\pi_0}\sqrt{1 - \frac{\pi_1}{\pi_0}x} + \sqrt{\pi_1}\sqrt{1-x}\right)^2, \quad x = p_{10}.$$

For $\pi_0 = 1/2$, efficiency is equal one for any x. If $\pi_0 \neq 1/2$, then the efficiency function attains a minimal value equal to 1, if and only if $x = \pi_0$, that corresponds to a chain corresponding to independent tests. Under a fixed $\pi_0 \neq 1/2$, the maximal efficiency for $x \in [\varepsilon_1, \min\{1, \pi_0/\pi_1\} - \varepsilon_2]$ is attained at one of the interval bounds and does not exceed 2, and the $R(h)$-optimum branching simulation design has the form

$$\beta^* = \left\{1, \sqrt{\frac{\pi_0}{\pi_1}(1-x)/(1 - \frac{\pi_1}{\pi_0}x)}\right\}.$$

Proof. Here $Q = 1 - p_{11} = x$. By the definition of $R_\beta(h)$, we have

$$\begin{aligned} R_\beta(h) &= \sum_{x=0}^{1} \pi(x)\,\beta(x) \sum_{x=0}^{1} \pi(x)\,\beta^{-1}(x)\,\mathcal{D}\tilde{\varphi}(x) = \\ &= h_{(0)}^T \left(I - Q^T\right)^{-1} B(\tau)(I-Q)^{-1} h_{(0)} = \\ &= \frac{h_1^2}{x^2}\left[\pi_0(p_{01} - p_{01}^2) + \beta_1^{-1}\pi_1(p_{11} - p_{11}^2)\right][\pi_0 + \beta_1\pi_1] = \\ &= \frac{h_1^2}{x}\pi_1\left[\left(1 - \frac{\pi_1}{\pi_0}x\right) + \beta_1^{-1}(1-x)\right][\pi_0 + \beta_1\pi_1]. \end{aligned}$$

The function $(a + b\beta)(c + d/\beta)$ is strictly convex with respect to β and attains its minimum for

$$\beta = \sqrt{da/bc}.$$

for any positive a, b, c, d. Therefore,

$$\beta^*(1) = (\pi_0/\pi_1)^{1/2} \left((1-x)\,/\,(1-(\pi_1/\pi_0)x)\right)^{1/2}$$

(denominator does not vanish, since $0 < x < \min\{1, \pi_0/\pi_1\}$).

Further,

$$R_{\beta^*}(h) = \frac{h_1^2}{x}\pi_1 \left(\pi_0(1-(\pi_1/\pi_0)x)^{1/2} + \sqrt{\pi_1(1-x)}\right)^2,$$

$$R_{\beta_0}(h) = \frac{h_1^2}{x}\pi_1 \left(2 - \frac{x}{\pi_0}\right),$$

and, hence,

$$\bar{R}(x,\pi) = \frac{R_{\beta_0}(h)}{R_{\beta^*}(h)} = (2 - x/\pi_0) \Big/ \left((\pi_0(1-(\pi_1/\pi_0)x))^{1/2} + (\pi_1(1-x))^{1/2}\right)^2.$$

Let us consider the function $f(x)$ under a fixed $\pi_0 \neq 1/2$,

$$f(x) = 1/\bar{R}(x,\pi_0).$$

Designate

$$f_1(x) = \left(\pi_0(1-\frac{\pi_1}{\pi_0}x)\right)^{1/2} + (\pi_1(1-x))^{1/2}, \quad f_2(x) = 2 - (x/\pi_0).$$

Compute a derivative $f'(x)$. We have $f(x) = f_1^2(x)/f_2(x)$,

$$f'(x) = \left[\frac{1}{\pi_0}\left(2 - \frac{x}{\pi_0}\right) - \frac{\sqrt{\pi_1}}{\sqrt{\pi_0}\left(1 - \frac{\pi_1}{\pi_0}x(1-x)\right)^{1/2}}\right] f(x).$$

Denote the expression in square brackets by $G(x)$. Since f_1 and f_2 do not vanish for $0 < x < \min\{1, \pi_0/\pi_1\}$, it is enough to investigate the behaviour of $G(x)$. Show that $G(x) = 0$ if and only if $x = \pi_0$. From $G(x) = 0$, it follows that

$$\sqrt{\pi_1\pi_0}\left(2 - \frac{x}{\pi_0}\right) = \left(\left(1 - \frac{\pi_1}{\pi_0}x\right)(1-x)\right)^{1/2}$$

Squaring both sides, we have

$$\pi_1\pi_0\left(2 - \frac{x}{\pi_0}\right)^2 = \left(1 - \frac{\pi_1}{\pi_0}x\right)(1-x),$$

whence $x = \pi_0$. Thus, the unique stationary point of $f(x)$ is $x = \pi_0$. Since, by definition, $f(x)$ does not exceed one and $f(\pi_0) = 1$, then $x = \pi_0$ is a maximum point. Therefore, the minimum of $f(x)$ is attained at one of the extreme points $x = \varepsilon_1$ or $x = \min\{1, \pi_0/\pi_1\} - \varepsilon_2$. Turning ε_1 and ε_2 to zero, we have

$$\min f(x)_{0<x<\min\{1,\pi_0/\pi_1\}} \geq \min\{f(0),\ f(\min\{1,\ \pi_0/\pi_1\})\} =$$
$$= \min\left\{\max\{\pi_0,\ \pi_1\},\ \frac{1+2\sqrt{\pi_0\pi_1}}{2}\right\} \geq \frac{1}{2}.$$

Thus, the efficiency maximum does not exceed 2 and is arbitrarily close to this value for $\varepsilon_i \to 0,\ \pi_0 \to 0$ or $\varepsilon_i \to 0,\ \pi_0 \to 1,\ i = 1, 2$.

The proof is complete.

10. Waiting and random walk processes

Define the random walk process $\{S_k;\ k \geq 0\}$ with a discrete state space $\Omega = \{0,\ \pm 1,\ \pm 2, \dots\}$ by the relations

$$S_0 = 0, \quad S_{k+1} = S_k + X_{k+1}, \quad k = 0, 1, \dots,$$

where $X_1, X_2, \dots$ are independent random values with the same distribution f, $f(k) = P\{X_1 = k\}$, $k \in \Omega$ is neither concentrated on the nonnegative nor on the nonpositive semiaxis.

Associate the waiting process $\{W_0, W_1, \dots\}$, defined by the relations

$$W_0 = 0, \quad W_{k+1} = \max\{0,\ W_k + X_{k+1}\}, \quad k = 0, 1, \dots, \tag{10.1}$$

where $X_1, X_2, \dots$ were defined earlier with the random walk process.

Both processes are Markov chains with a discrete state space (Ω for the random walk and $\mathcal{X}' = \{0, 1, 2, \dots\}$ for the waiting process). Here elements of the transition matrix have the form

$$p(x, y) = f(y - x), \quad x, y \in \Omega$$

for the random walk and

$$p(x, y) = f(y - x), \quad x, y \in \mathcal{X}', \quad y \neq 0,$$
$$p(x, 0) = \sum_{z=-\infty}^{0} f(z - x)$$

for the waiting process.

Designate

$$M_n = \max_{0 \leq k \leq n} S_k,$$
$$M_n(y) = P\{M_n < y\},$$
$$W_n(y) = P\{W_n < y\}.$$

for $y \geq 0$.

Further, the random walk theory results stated in [40, vol. 2] are used.

It is known that the random variables M_n and W_n have the same distribution function ($W_n(y) = M_n(y)$, $n = 1, 2, \dots$) and, respectively, converge in distribution to random variates M and W, having the same distribution functions, for $EX_1 < 0$. Denote these functions by $M(y)$. We have

$$M(y) = P\{M < y\},$$

where $M = \sup_{0 \leq k < \infty} S_k$.

One of the main problems for a random walk is finding a value,

$$\theta = \bar{M}(x_0) = 1 - M(x_0) = P\{M \geq x_0\},$$

which coincides with $P\{W \geq x_0\}$ – the excess probability of a given value x_0 by the value of waiting time in the stationary limit. In the general case, the value θ has no closed analytical representation and may be evaluated with the help of statistical simulation.

Consider a Markov chain $\{W_0, W_1, \dots\}$. It has a discrete state space $\mathcal{X}' = \{0, 1, \dots\}$. The chain $\{W_k;\ k \geq 0\}$ has the stationary distribution $\pi = \{\pi(0),\ \pi(1), \dots\}$ for $EX_1 < 0$. Functional evaluating is required:

$$\theta = J_{st}(h) = \sum_{x=0}^{\infty} \pi(x)\, h(x) = P\{W \geq x_0\},$$

where $h(x) = h_{x_0}(x) = 0,\ \ x < x_0,\ \ h(x) = h_{x_0}(x) = 1,\ \ x \geq x_0$. Identifying $\xi_k = W_k$, we will use the estimator $\tilde{J}_\beta(h; m)$ defined in Section 4. Let

$$\begin{aligned}
&\tilde{h}(x) = h(x) - \theta, \\
&\tilde{\varphi}(x, 0) = \tilde{h}(x), \\
&\tilde{\varphi}(x, k+1) = \sum_{y \in \mathcal{X}} f(y - x)\, \tilde{\varphi}(y, k) + \tilde{h}(x), \quad k \geq 0, \\
&\tilde{h}(x) = \tilde{h}_{x_0}(x) = h_{x_0}(x) - \theta, \quad x \in \mathcal{X}.
\end{aligned}$$

The following lemma is valid.

Lemma 10.1. *For the equation*

$$\tilde{\varphi}(x) = \sum_{y \in \mathcal{X}} f(y - x)\, \tilde{\varphi}(y) + \tilde{h}(x), \tag{10.2}$$

there exists a solution $\tilde{\varphi}(x)$ *satisfying the conditions:* $\tilde{\varphi}(x) < \infty$, $x \in \mathcal{X}$ *and* $\lim_{k \to \infty} \tilde{\varphi}(x, k) = \tilde{\varphi}(x)$.

Designate

$$D_p\tilde{\varphi}(x) = \sum_{y=1}^{\infty} f(y-x)\,\tilde{\varphi}^2(y) - \left(\sum_{y=1}^{\infty} f(y-x)\,\tilde{\varphi}(y)\right)^2,$$

$$r_p(\tilde{\varphi}, x) = \sum_{y=1}^{\infty} \gamma(x,y)\, f(y-x)\, [\tilde{\varphi}(y)\,\beta(x)/\beta(y)],$$

where $\gamma(x,y) = \overline{\beta(y)/\beta(x)} - (\overline{\beta(y)/\beta(x)})^2$,

$$\beta(0) = 1, \quad \beta(x) > 0, \quad x \in \mathcal{X}, \quad \sum_{0}^{\infty} \pi(x)\,\beta(x) < \infty,$$

$$\tau(x) = \beta(x)\,\pi(x) \,/ \sum_{x \in \mathcal{X}} \pi(x)\,\beta(x).$$

Then Theorem 4.1 may be reformulated as follows.

Theorem 10.1. *Let $\{W_k;\, k \geq 0\}$ be a waiting process defined by formula (10.1) and $EX_1 < 0$, $\sum_{x=0}^{\infty} \pi(x)\,\beta(x) < \infty$. Then the quantity $\tilde{J}_\beta(h,m)$ defined by formula (4.3) is asymptotically normal and an asymptotically unbiased estimator of the quantity $\theta = P\{W \geq x_0\}$ for $h = h_{x_0}$, where $x_0 > 0$.*
Here

$$E\tilde{T}_\beta \lim_{m\to\infty} m^{-1}\mathcal{D}\tilde{J}_\beta(h;m) = \sum_{0}^{\infty} \pi^2(x)\,\tau^{-1}(x)\,[\mathcal{D}_p\tilde{\varphi}(x) + r_p(\tilde{\varphi}, x)],$$

where $\tau(x) = \beta(x)\,\pi(x) \,/ \sum_0^\infty \pi(y)\,\beta(y)$.

By virtue of Theorem 4.1, it suffices to prove Lemma 10.1 to prove Theorem 10.1.

Proof. First let us prove the existence of the equation solution

$$\varphi(x) = \sum_{y=1}^{\infty} f(y-x)\,\varphi(y) + h(x),$$

where $h(x) = h_{x_0}(x)$,to be such that $\varphi(x) < \infty$, $x \in \mathcal{X}$ and $\varphi(x,k) \to \varphi(x)$ for $k \to \infty$ where $\varphi(x,k)$ is defined in (2.6) for $Q_k(x,y) = Q(x,y) = f(y-x)$, $h(x,k) = h_{x_0}(x)$, $k = 1, 2, \ldots$

Designate random values

$$\bar{\varphi}(x,k) = \sum_{i=1}^{k} h(W_i), \quad \bar{\varphi}(x) = \sum_{i=1}^{\alpha} h(W_i),$$

provided that $W_1 = x$, $\bar{k} = \min\{k, \alpha\}$, $\alpha = \{l : W_2, \dots, W_{l-1} > 0, W_l = 0\}$. We have $\bar{\varphi}(x, 0) = h(x)$

$$\bar{\varphi}(x,k) = h(x) + \sum_{y=1}^{\infty} \zeta(x,y)\,\bar{\varphi}(y,k-1), \quad k \geq 1,$$

where $\zeta(x,y)$ is a random value such that

$$\zeta(x,y) = \begin{cases} 1, & \text{if } W_2 = y > 0, \\ 0, & \text{if } W_2 = 0. \end{cases}$$

It is evident that

$$E\zeta(x,y) = f(y-x), \quad E\zeta^2(x,y) = f(y-x),$$
$$E\zeta(x,y)\zeta(x,z) = 0, \quad y \neq z.$$

Note that the decreasing ladder process is the proper process for $EX_1 < 0$, the first lower ladder index has a finite expectation ($E\tau_1 < \infty$), and the average number of lower weak ladder points in the interval $(0, x)$ is finite ($E\bar{H}^-(x) < \infty$). Hence,

$$E\alpha = E\tau_1 E\bar{H}^-(x) < \infty.$$

Therefore, with probability equal to 1, $\alpha < \infty$ and

$$\lim_{k \to \infty} \bar{\varphi}(x,k) = \bar{\varphi}(x).$$

Hence,

$$\lim \varphi(x,k) = E\bar{\varphi}(x) < \infty.$$

Designate $\varphi(x) = E\bar{\varphi}(x)$. Then we have that $\varphi(x)$ satisfies the condition

$$\varphi(x) = \sum_{y=1}^{\infty} f(y-x)\,\varphi(y) + h(x). \tag{10.3}$$

Let $\varphi_0(x,k) = \varphi(x,k)$ for $x_0 = 0$. According to the previous reasoning, $\varphi_0(x,k) \to \varphi_0(x)$ for $k \to \infty$, $\varphi_0 < \infty$ and $\varphi_0(x)$ is the solution of (10.3) with $x_0 = 0$. Note that

$$\tilde{\varphi}(x,k) = \varphi(x,k) - \theta\varphi_0(x,k)$$

and

$$\tilde{\varphi}(x,k) \to \varphi(x) - \theta\varphi_0(x),$$

for $k \to \infty$, where $\varphi(x) - \theta\varphi_0(x) < \infty$.

Designating $\tilde{\varphi}(x) = \varphi(x) - \theta\varphi_0(x)$, we have that $\tilde{\varphi}(x)$ is the solution of (10.2) with necessary characteristics. Q.E.D.

For the waiting process, we can find designs in an explict form which are close to $R(h)$-optimum designs for large values x_0. Precisely, the next results are valid.

Theorem 10.2. *Consider the waiting process described earlier. Suppose the following conditions are fulfilled:* $EX_1 < 0$

$$\sum_{x=-\infty}^{\infty} f(x)\, e^{\lambda x} = 1 \tag{10.4}$$

for some $\lambda > 0$,

$$f(x) \leq Ce^{-\alpha x}, \quad x > 0.$$

for some $\alpha > \lambda$.

If $\beta(x) = \beta^*(x) = e^{-\gamma x}$, $x \leq x_0$, $\gamma = \min(\lambda, \alpha/2)$, $\beta(x) = \beta^*(x_0)$, $x > x_0$, *then for* $x_0 \to \infty$, *in the case of* $\alpha \geq 2\lambda$ *we have* $\theta \to 0$

$$\mathcal{D}\hat{\theta}_\beta E\widehat{T}_\beta = O(\theta^2 \ln^2(1/\theta))$$

or else $(\alpha < 2\lambda)$

$$\mathcal{D}\hat{\theta}_\beta E\widehat{T}_\beta = O(\theta^{\alpha/\lambda}).$$

Theorem 10.3. *If, in Theorem 10.2, function* $\beta^*(x)$ *is replaced by*

$$\bar{\beta}(x) = e^{\gamma_1 x}, \quad x \leq x_0, \quad \bar{\beta}(x) = \bar{\beta}(x_0), \quad x > x_0,$$

where $0 \leq \gamma_1 \leq 2\gamma$, *then*

$$\mathcal{D}\hat{\theta}_\beta E\widehat{T}_\beta = O(\theta^v),$$

where $v = 2 - \left|\frac{\gamma_1 - \lambda}{\lambda}\right|$ *for* $\gamma_1 \neq \lambda$, $\alpha \geq 2\lambda$ *and for* $\alpha < 2\lambda$,

$$v = \begin{cases} \frac{2\gamma}{\lambda} - \frac{\gamma_1 - \lambda}{\lambda}, & \gamma_1 \geq \lambda, \\ \frac{2\gamma}{\lambda} - \frac{2\gamma - \gamma_1 - \lambda}{\lambda}, & \gamma_1 \leq 2\gamma - \lambda, \\ \frac{2\gamma}{\lambda}, & 2\gamma - \lambda < \gamma_1 < \lambda. \end{cases}$$

From Theorem 10.3, it follows that the design $\bar{\beta}(x)$ is more efficient than $\beta_0(x) \equiv 1$ for $2\lambda > \gamma_1 > \lambda/2$, i.e., this theorem characterizes the stability of design $\bar{\beta}(x)$ with respect to an error in λ determination.

The following lemma is necessary for the demonstration of Theorems 10.2 and 10.3.

Lemma 10.2. *If for some* $\lambda > 0$,

$$\sum_{x=-\infty}^{\infty} f(x)\, e^{\lambda x} = 1,$$

then

$$\varphi(x) \sim \begin{cases} c_1 e^{\lambda(x - x_0)}, & x \leq x_0, \quad x, x_0 \to \infty, \\ c_1 + c_2(x - x_0), & x > x_0, \quad x, x_0 \to \infty, \end{cases}$$

where c_1, c_2 are some positive constants.

Proof of Lemma 10.2. From the hypothesis of the lemma, it follows that $\pi(y) \sim \text{const}\, e^{-\lambda y}$. Let x be a number greater than 0. Consider the behaviour of the chain $\{W_k;\ k \geq 1\}$ before the first hit into the set $[0, x)$ for $W_1 = x > 0$. It coincides with the behavior of a random walk $\{S_k;\ k \geq 1\}$ for $S_1 = x$ before the first hit into the set $(-\infty, x)$. From the random walk theory, it is known that the number of chain steps until the event described above is finite with probability 1 and has a finite expectation under the condition of $EX_1 < 0$. Denote this number by γ. We have $E\gamma < \infty$. Denote by $\bar{\Pi}(x_0, x)$ the chain $\{W_k;\ k \geq 1\}$, $W_1 = x$ number of hits into $[x_0, \infty)$ before the first hit into $[0, x)$. Let $\Pi(x_0, x) = E\bar{\Pi}(x_0, x)$. By the duality lemma [40, v.2], this number is equal to $\Psi(\infty) - \Psi(x^*)$, where $x^* = \max\{x, x_0\}$ and $\Psi(x) = M(x)/\Psi(\infty)$ is the mean number of strict upper ladder points in the interval $[0, x]$, $M(x) = \sum_{y=0}^{x} \pi(y)$.

Let $\bar{\varphi}(x)$ be the number of hits into (x_0, ∞) of the chain $\{S_k;\ k \geq 1\}$, $S_1 = x$ before the first hit into $(-\infty, 0]$. Then

$$\bar{\varphi}(x) = \bar{\Pi}(x_0, x) + \bar{\varphi}(y), \tag{10.5}$$

where y is a random variate with distribution $\rho(y - x)$. Here

$$\rho(z) = \sum_{n=1}^{\infty} P\{W_1 > 0,\ W_2 > 0, \ldots, W_{n-1} > 0, W_n = z < 0\}.$$

Note that $\bar{\varphi}(x)$ satisfies the equation

$$\bar{\varphi}(x) = \sum_{y=1}^{\infty} \zeta(x, y)\, \bar{\varphi}(y) + h(x),$$

where

$$\zeta(x, y) = \begin{cases} 1 & \text{with probability } f(y - x), \\ 0 & \text{with probability } 1 - f(y - x). \end{cases}$$

As it is shown in Lemma 10.1, $\varphi(x) = E\bar{\varphi}(x)$ is an iterative solution of the equation

$$\varphi(x) = \sum_{y=1}^{\infty} f(y - x)\, \varphi(y) + h(x).$$

On the other hand, computing the expectation of the both sides of (10.5), we have

$$\varphi(x) = Z(x) + \sum_{y=0}^{x} \rho(y - x)\, \varphi(y), \tag{10.6}$$

where the supposition of $\varphi(0) = 0$, $\rho(0) = 0$ and $Z(x) = \Pi(x, x_0)$ has been admitted.

Equation (10.6) is a renewal equation where the function $Z(x)$ is bounded. Therefore, the unique solution bounded in the intervals (only such a solution can be the $\bar{\varphi}(x)$ expectation) is [40, v. 2]

$$\varphi(x) = \sum_{y=0}^{x} \bar{H}'(x-y)\, Z(y),$$

where $\bar{H}'$ is a renewal function corresponding to the lower ladder process. Since this process is proper for $EX_1 < 0$, then by the renewal theorem, $\bar{H}'(y) \sim$ const for $y \to \infty$. Moreover,

$$Z(x) \sim \begin{cases} e^{\lambda(x-x_0)}\text{const}, & x < x_0, \quad x, x_0 \to \infty, \\ \text{const}, & x \geq x_0, \quad x, x_0 \to \infty, \end{cases}$$

since $Z(x) = \Psi(\infty) - \Psi(x^*)$ and $\Psi(x) \sim e^{-\lambda x}$const.

Hence

$$\varphi(x) \sim \begin{cases} c_1 e^{\lambda(x-x_0)}, & x < x_0, \quad x, x_0 \to \infty, \\ c_1 + c_2(x - x_0), & x \geq x_0, \quad x, x_0 \to \infty, \end{cases}$$

Q.E.D.

Assume $\varphi_0(x) = \varphi(x)$ for $h(x) = \bar{h}(x) \equiv 1$. Then

$$\tilde{\varphi}(x) = \varphi(x) - \theta\varphi_0(x).$$

Since $\varphi_0(x) = \varphi(x)$ for $x_0 = 0$ then by lemma 10.2 $\varphi_0(x) \sim c_2 x, \quad x \to \infty$ (if condition (10.4) is fulfilled).

Lemma 10.3. *If the conditions of Lemma 10.2 are valid and $f(x) \leq Ce^{-\alpha x}$, where $\alpha > \lambda$, C is a constant, then*

$$\mathcal{D}_p\tilde{\varphi}(x) = \begin{cases} O\left(e^{2\gamma(x-x_0)}\right), & x \leq x_0, \quad x, x_0 \to \infty, \\ O(1), & x > x_0, \quad x, x_0 \to \infty, \end{cases}$$

where $\gamma = \min(\alpha/2, \lambda)$. Moreover, if $\beta(x) = \beta^*(x)$, then

$$r_p(\tilde{\varphi}, x) = \begin{cases} O\left(e^{2\gamma(x-x_0)}\right), & x \leq x_0, \quad x, x_0 \to \infty, \\ O\left((x-x_0)^2\right), & x > x_0, \quad x, x_0 \to \infty, \end{cases}$$

Proof of Lemma 10.3. First consider the quantity $\mathcal{D}_p\varphi(x)$. Present it as the sum

$$\mathcal{D}_p\varphi(x) = I_1(x) + I_2(x),$$

where

$$I_1(x) = \sum_{y=1}^{x_0} f(y-x)\varphi^2(y),$$

$$I_2(x) = \sum_{y=x_0+1}^{\infty} f(y-x)\varphi^2(y) - \left(\sum_{y=1}^{\infty} f(y-x)\varphi(y)\right)^2.$$

Investigate $I_1(x)$ for $x \le x_0$. Let $F(x) = \sum_{y=-\infty}^{x} f(y)$.

According to Lemma 10.2, we have

$$I_1(x) = C_1 \sum_{y=1}^{x_0} f(y-x)\, e^{2\lambda(y-x_0)} =$$

$$= C_1 \sum_{y=1}^{x} f(y-x)\, e^{2\lambda(y-x_0)} + C_1 \sum_{y=x+1}^{x_0} f(y-x)\, e^{2\lambda(y-x_0)}.$$

The first term does not exceed $C_1 F(0)\, e^{2\lambda(x-x_0)}$ and we have the following estimator for the second one:

$$CC_1 \sum_{y=x+1}^{x_0} e^{-\alpha(y-x)} e^{2\lambda(y-x_0)} \le \text{const}\, e^{2\gamma(x-x_0)}.$$

For $x > x_0$, we have

$$I_1(x) \le C_1 \sum_{y=1}^{x_0} f(y-x)\, e^{2\lambda(y-x_0)} \le C_1 F(0) \sum_{y=1}^{x_0} e^{2\lambda(y-x_0)} = \mathrm{O}(1).$$

Thus,

$$I_1(x) = \begin{cases} \mathrm{O}\left(e^{2\gamma(x-x_0)}\right), & x \le x_0, \quad x, x_0 \to \infty, \\ \mathrm{O}(1), & x > x_0, \quad x, x_0 \to \infty. \end{cases}$$

We now investigate $I_2(x)$. Let $x \le x_0$. We have

$$I_2(x) \le \sum_{y=x_0+1}^{\infty} f(y-x)\, \varphi^2(y)$$

and, according to Lemma 10.2,

$$I_2(x) \le \sum_{y=x_0+1}^{\infty} f(y-x)\, \left[(y-x_0)C_2 + C_1\right]^2 =$$

$$= C_1^2 \sum_{y=x_0+1}^{\infty} f(y-x) + 2C_1C_2 \sum_{y=x_0+1}^{\infty} f(y-x)\,(y-x_0) +$$

$$+ C_2^2 \sum_{y=x_0+1}^{\infty} f(y-x)\,(y-x_0)^2 \le$$

$$\le \text{const} \sum_{y=x_0+1}^{\infty} f(y-x)\,(y-x_0)^2 \le$$

$$\le \text{const}\, e^{\alpha(x-x_0)} \sum_{y=1}^{\infty} e^{-\alpha(y-x_0)}(y-x_0)^2 = \mathrm{O}\left(e^{\alpha(x-x_0)}\right).$$

For $x > x_0$, we have

$$I_2(x) \le \text{const} \sum_{y=x_0+1}^{\infty} f(y-x)\,(y-x_0)^2.$$

Decompose the sum from the right side into two sums: from x_0+1 to x and from $x+1$ to ∞. The first one is denoted by $I_2'(x)$ and the second by $I_2''(x)$.

By virtue of the assumptions of the lemma $EX_1^2 < \infty$. Therefore,

$$\begin{aligned} I_2'(x) &= \sum_{y=x_0+1}^{x} f(y-x)\,(y-x)^2 + 2\sum_{y=x_0+1}^{x} f(y-x)\,(y-x)\,(x-x_0) + \\ &+ \sum_{y=x_0+1}^{x} f(y-x)\,(x-x_0)^2 \le EX_1^2 + 2(x-x_0)EX_1 + 2(x-x_0)^2 = \\ &= \mathrm{O}\left((x-x_0)^2\right). \end{aligned}$$

By assumption, $f(y-x) \le Ce^{-\alpha(y-x)}$ for $y-x>0$. Therefore,

$$\begin{aligned} I_2''(x) \le C\Bigg\{ &\sum_{y=x+1}^{\infty} e^{-\alpha(y-x)}\,(y-x)^2 + \sum_{y=x+1}^{\infty} e^{-\alpha(y-x)}(x-x_0)^2 + \\ &+2\sum_{y=x+1}^{\infty} e^{-\alpha(y-x)}(y-x)\,(x-x_0)\Bigg\} = \mathrm{O}\left((x-x_0)^2\right). \end{aligned}$$

Thus

$$\mathcal{D}\varphi(x) = \begin{cases} \mathrm{O}\left(e^{2\gamma(x-x_0)}\right), & x \le x_0, \quad x, x_0 \to \infty, \\ \mathrm{O}\left((x-x_0)^2\right), & x > x_0, \quad x, x_0 \to \infty. \end{cases}$$

The estimator for $\mathcal{D}\tilde{\varphi}(x)$ has the same form. This can be verified by a similar technique.

Let us consider the quantity $r_p(\varphi, x)$. Represent $r_p(\varphi, x)$ as

$$r_p(\varphi, x) = I_1(x) + I_2(x),$$

where

$$I_1(x) = \sum_{y=1}^{x_0} f(y-x)\,\gamma(x,y)\,[\varphi(y)\,\beta(x)/\beta(y)]^2 \le \frac{1}{4}\sum_{y=1}^{x_0} f(y-x)\,[\varphi(y)\,\beta(x)/\beta(y)]^2,$$

$$\begin{aligned} I_2(x) &= \sum_{x_0+1}^{\infty} \gamma(x,y)\,f(y-x)\,[\varphi(y)\,\beta(x)/\beta(y)]^2 \le \\ &\le \frac{1}{4}\sum_{y=x}^{\infty} f(y-x)\,[\varphi(y)\,\beta(x)/\beta(y)]^2 + \frac{1}{4}\sum_{y=1}^{x-1} f(y-x)\,\varphi^2(y)\,e^{2\gamma(x-y)}. \end{aligned}$$

Let $\beta(x) = \beta^*(x) = e^{\gamma x}$, $x \le x_0$, $\beta(x) = \beta^*(x_0)$, $x > x_0$. Then we have

$$I_1(x) \le \frac{1}{4}\sum_{y=x}^{x_0} f(y-x)\,\varphi^2(y)\,e^{2\gamma(x-y)} + \frac{1}{4}\sum_{y=1}^{x-1} f(y-x)\,\varphi^2(y)\,e^{2\gamma(x-y)}.$$

for $x \le x_0$ The first sum does not exceed

$$\begin{aligned}
&\text{const}\sum_{y=x}^{x_0} e^{-\alpha(y-x)} e^{2\lambda(y-x_0)} e^{2\gamma(x-y)} = \\
&= \text{const}\sum_{y=x}^{x_0} e^{(-\alpha+2\lambda-2\gamma)y+(\alpha+2\gamma)x-2\lambda x_0} \le \\
&\le \text{const}\, e^{2\lambda(x-x_0)} \le \text{const}\, e^{2\gamma(x-x_0)},
\end{aligned}$$

since $-\alpha+2\lambda-2\gamma = -\alpha < 0$ if $\alpha > 2\lambda$ (here $2\gamma = 2\lambda$) and $-\alpha+2\lambda-2\gamma = 2(\lambda-\alpha) < 0$ if $\alpha \le 2\lambda$ (here $2\gamma = \alpha$). The second sum does not exceed

$$\text{const}\,\frac{1}{4}\,F(0)\sum_{y=1}^{x-1} e^{2\lambda(y-x_0)+2\gamma(x-y)} \le \text{const}\, e^{2\lambda(x-x_0)} \le \text{const}\, e^{2\gamma(x-x_0)},$$

since $2\gamma = \min\{\alpha, 2\lambda\} \le 2\lambda$, $x - x_0 \le 0$.

Thus, for $x \le x_0$,

$$I_1(x) = \mathrm{O}\left(e^{2\gamma(x-x_0)}\right).$$

For $x > x_0$,

$$\begin{aligned}
I_1(x) &\le \frac{1}{4}\sum_{y=1}^{x_0} f(y-x)\,\varphi^2(y)\,e^{2\gamma(x_0-y)} \le \\
&\le \text{const}\sum_{y=1}^{x_0} f(y-x)\,e^{(y-x_0)(2\lambda-2\gamma)} \le \text{const} = \mathrm{O}(1).
\end{aligned}$$

For $I_2(x)$, by Lemma 10.2 we have

$$\begin{aligned}
I_2(x) &\le \frac{1}{4}C\sum_{y=x_0+1}^{\infty} f(y-x)\,(y-x_0)^2 e^{2\gamma(x-x_0)} \le \\
&\le \frac{1}{4}C\sum_{y=x_0+1}^{\infty} f(y-x)\,(y-x)^2 e^{2\gamma(x-x_0)} \le \frac{1}{4}C\,EX_1^2 e^{2\gamma(x-x_0)} = \\
&= \mathrm{O}\left(e^{2\gamma(x-x_0)}\right).
\end{aligned}$$

for $x \le x_0$. For $x > x_0$, $\beta^*(x) = e^{\gamma x_0}$ and

$$I_2(x) \le C \sum_{y=x_0+1}^{x} f(y-x)\,[(y-x_0)C_1 + C_2]^2 + \sum_{y=x_0+1}^{\infty} f(y-x)\,[(y-x_0)C_1 + C_2]^2.$$

This expression coincides with the estimator $I_2(x)$ in the expansion $\mathcal{D}_p\varphi(x)$ and, hence, has the form $\mathrm{O}\left((x-x_0)^2\right)$.

Thus,

$$r_p(\varphi, x) = \begin{cases} \mathrm{O}\left(e^{2\gamma(x-x_0)}\right), & x \le x_0, \quad x, x_0 \to \infty, \\ \mathrm{O}\left((x-x_0)^2\right), & x > x_0, \quad x, x_0 \to \infty. \end{cases}$$

Calculations for $r_p(\tilde{\varphi}, x)$ are carried out in the same way since

$$r_p(\tilde{\varphi}, x) = r_p(\varphi, x) + r_p(\varphi_0, x)\theta^2 - 2\theta \sum f(y-x)\,\frac{\varphi(y)\,\varphi_0(y)}{\beta^2(y)}\,\beta^2(x)\gamma(x,y),$$

$\theta \sim Ce^{-\lambda x_0}$ for $x \to x_0$ and gives the same result.

Thus, Lemma 10.3 is complete.

Now let us prove Theorem 10.2. By virtue of Theorem 10.1 and Lemma 10.3, we have

$$\mathcal{D}\hat{\theta}_\beta E\widehat{T}_\beta = \left(\sum_{x=0}^{x_0} \pi(x)\,\beta^{*-1}(x)\;\mathrm{O}\left(e^{2\gamma(x-x_0)}\right) + \beta^{*-1}(x_0) \sum_{x=x_0+1}^{\infty} \pi(x)\;\mathrm{O}\left((x-x_0)^2\right)\right) \times$$
$$\times \left[\sum_{x=0}^{x_0} \pi(x)\,\beta^*(x) + \beta^*(x_0) \sum_{x=x_0+1}^{\infty} \pi(x)\right], \tag{10.7}$$

for $\beta(x) = \beta^*(x)$. Here

$$\pi(x) \sim \mathrm{const}\, e^{-\lambda x}, \quad \theta \sim \mathrm{const}\, e^{-\lambda x_0} \quad (x \to \infty,\; x_0 \to \infty).$$

Moreover,

$$\sum_{x=0}^{x_0} e^{(\gamma-\lambda)x} = x_0 + 1 + \mathrm{O}(x_0) = \mathrm{O}(\ln(1/\theta)) \quad \text{for } \gamma = \lambda,\; x_0 \to \infty,$$

$$\sum_{x=0}^{x_0} e^{(\gamma-\lambda)x} < \sum_{x=0}^{\infty} e^{(\gamma-\lambda)x} = \mathrm{O}(1) \quad \text{for } \gamma < \lambda,$$

$$\sum_{x=x_0+1}^{\infty} e^{-\lambda x} = e^{-\lambda x_0}\,\mathrm{O}(1), \quad x_0 \to \infty.$$

Substituting the representations for $\pi(x)$ and $\beta^*(x)$ into formula (10.7), we have

$$\mathcal{D}\hat{\theta}_\beta E\widehat{T}_\beta = \mathrm{O}\left(e^{-2\gamma x_0}\right)\left[\mathrm{O}\left(\sum_{x=0}^{x_0} e^{(\gamma-\lambda)x}\right) + e^{(\gamma-\lambda)x_0}\,\mathrm{O}(1)\right]^2,$$

for $x_0 \to \infty$. For $\alpha \geq 2\lambda$, we have $\gamma = \lambda$ and

$$\mathcal{D}\hat{\theta}_\beta E\widehat{T}_\beta = \mathrm{O}\left(\theta^2 \ln^2(1/\theta)\right) \quad (x_0 \to \infty).$$

For $\alpha < 2\lambda$, we have $\gamma = \alpha/2, \;\; \gamma < \lambda$

$$\mathcal{D}\hat{\theta}_\beta E\widehat{T}_\beta = \mathrm{O}\left(e^{-2\gamma x_0}\right) = \mathrm{O}\left(\theta^{\alpha/\lambda}\right).$$

The proof of Theorem 10.2 is now complete.

Proof of Theorem 10.3. Replace $\beta^*(x)$ by $\bar{\beta}(x) = e^{-\gamma_1 x}, \;\; x < x_0, \;\; \bar{\beta}(x_0), \;\; x > x_0$ in formula (10.7). We have

$$\mathcal{D}\hat{\theta}_\beta E\widehat{T}_\beta =$$

$$= \mathrm{O}\left(e^{-2\gamma x_0}\right)\left[\mathrm{O}\left(\sum_{x=0}^{x_0} e^{(2\gamma-\gamma_1-\lambda/x)x}\right) + e^{(2\gamma-\gamma_1-\lambda)x_0}\,\mathrm{O}(1)\right] \times$$

$$\times \left[\mathrm{O}\left(\sum_{x=0}^{x_0} e^{(\gamma_1-\lambda/x)x}\right) + e^{(\gamma-\lambda)x_0}\,\mathrm{O}(1)\right].$$

Hence, in the case of $\alpha \geq 2\lambda$, we have $\gamma = \lambda$ and

$$\mathcal{D}\hat{\theta}_\beta E\widehat{T}_\beta = \mathrm{O}\left(\theta^{2-\left|\frac{\gamma_1-\lambda}{\lambda}\right|}\right), \quad \gamma_1 \neq \lambda,$$

since for $\gamma_1 < \lambda$, the second multiplier has the form of $\mathrm{O}(1)$, the third one – $\mathrm{O}\left(e^{(\gamma_1-\lambda)x_0}\right)$, and for $\gamma_1 > \lambda$ the second multiplier has the form of $\mathrm{O}\left(e^{(\lambda-\gamma_1)x_0}\right)$, and the third one – $\mathrm{O}(1)$. The case of $\gamma_1 = \lambda$ is considered in Theorem 10.2.

If $\alpha < 2\lambda$ we have $\gamma = \alpha/2 < \lambda$. Let $\gamma_1 > \lambda$. Here the second multiplier has the form of $\mathrm{O}(1)$ and the third one – $\mathrm{O}\left(e^{(\gamma_1-\lambda)x_0}\right)$. Hence,

$$\mathcal{D}\hat{\theta}_\beta E\widehat{T}_\beta = \mathrm{O}\left(\theta^{\frac{\alpha}{\lambda}-\frac{\gamma_1-\lambda}{\lambda}}\right).$$

If $\lambda > \gamma_1 \geq 2\gamma - \lambda$, then the second and third multipliers have the form $\mathrm{O}(1)$ and

$$\mathcal{D}\hat{\theta}_\beta E\widehat{T}_\beta = \mathrm{O}\left(\theta^{\alpha/\lambda}\right).$$

Finally, if $\gamma_1 < 2\gamma - \lambda$, then the second multiplier has the form $O\left(e^{(2\gamma-\gamma_1-\lambda)x_0}\right)$, the third one – $O(1)$ and

$$\mathcal{D}\hat{\theta}_\beta E\widehat{T}_\beta = O\left(e^{-\gamma_1-\lambda}\right) = O\left(\theta^{1+\gamma_1/\lambda}\right).$$

Thus

$$\mathcal{D}\hat{\theta}_\beta E\widehat{T}_\beta = O\left(\theta^v\right),$$

where

$$v = \begin{cases} \frac{2\gamma}{\lambda} - \frac{\gamma_1-\lambda}{\lambda}, & \gamma_1 \geq \lambda, \\ \frac{2\gamma}{\lambda} - \frac{2\gamma-\gamma_1-\lambda}{\lambda}, & \gamma_1 \leq 2\gamma-\lambda, \\ \frac{2\gamma}{\lambda}, & 2\gamma-\lambda < \gamma_1 < \lambda. \end{cases}$$

Q.E.D.

11. The rank lattice chain

In this section, we will investigate regular Markov chains with the finite state space $\{0, 1, \ldots, n+1\}$ which have transition "upwards"probabilities ($p_{i,i+k}$ for $k > 0$) different from zero only for the near state ($p_{i,i+k} = 0,\ k \geq 2$). Those chains can be applied to the investigation of the reliability of a complex system [66]. The main characteristic to be evaluated is the hitting into a state $n+1$ probability before returning into 0.

If we take $p_{n+10} = 1$, the above-mentioned probability remains the same. Therefore, we will consider finite regular Markov chains with the state space $\{0, 1, \ldots, n+1\}$, transition matrix $P = (p_{ij})_{i,j=0}^{n+1}$ and the relations $p_{i,j+k} = 0,\ k \geq 2,\ p_{n+1,0} = 1$. Let us also consider that the states are enumerated so that $\pi_0 \geq \pi_1 \geq \ldots \geq \pi_{n+1}$.

The present section is devoted to some additional results, which can be obtained for this class of chains.

Let $\xi_0 = 0$. Let us denote the random variates connected with the behaviour of $\{\xi_k\}$ before returning into the state $\{0\}$ by letters with a bar and their expectations by the same letters without a bar. Let $\bar{\nu}_i$ be the number of state i attainments from below, $\bar{\omega}_{ij}$ be the number of state j attainments under the condition $\xi_0 = i$, and $\bar{T}$ be the steps amount until returning to zero.

Let $h_j(x) = 0$ for $x \neq j$ for some fixed $j,\ j = 0, 1, \ldots, n+1,\ h_j(j) = 1$.

Consider branching simulation designs of the form $1 = \beta_0 \leq \beta_1 \leq \ldots \leq \beta_{n+1}$, where β_{i+1}/β_i are intergers. Set

$$\tilde{\bar{\omega}}_{0i} = \widehat{J}_\beta(h_i),$$

where $\widehat{J}_\beta(h)$ is defined by formula (4.3).

We will call the corresponding technique splitting and roulette to distinguish it from one developed below in Section 11.1.

Designate $\gamma_k = P\{\bar{\nu}_k \geq 1\},\ \alpha_k = P\{\bar{\nu}_k \geq 2\}/\gamma_k,\ k = 1, 2, \ldots, n,\ \alpha_0 = \alpha_{n+1} = \alpha_1 = 0,$ $\gamma_0 = 1$.

Theorem 11.1. *Let $\{\xi_k;\ k \geq 0\}$ be a regular Markov chain with the state space $\{0,1,\dots,n+1\}$ and transition matrix P, where $p_{i,i+s} = 0,\ \ s \geq 2,\ \ p_{n+10} = 1$. Then the formulas*

$$\mathcal{D}\tilde{\bar{\omega}}_{0,n+1} = \pi_{n+1}^2 \sum_{k=0}^{n} \beta_k^{-1} \left[\gamma_{k+1}^{-1}(1-\alpha_{k+1})^{-1} - \gamma_k^{-1}(1-\alpha_k)^{-1}\right].$$

are valid.

Proof. From Theorem 3.1 and Remark 3.2, we have

$$\mathcal{D}\tilde{\bar{\omega}}_{0,n+1} = \mathcal{D}\widehat{J}_\beta(h_i) =$$
$$= \sum_{k=0}^{n+1} \beta_k^{-1} \left\{\omega_{0k}\left[2\omega_{kk}\delta_{kj} - \delta_{kj} - \omega_{kj}^2\right] + \sum_{l=1}^{n+1} p_{kl}\omega_{lj}^2 + \sum_{l=n+1}^{n+1} \omega_{0l}p_{kl}\omega_{kj}^2\right\}.$$

Designate $Q = (p_{ij})_{i,j=1}^{n+1}$. Note that $(\omega_{ij})_{i,j=1}^{n+1} = (1-Q)^{-1}$.
Let $\{\xi_k';\ k \geq 0\}$ be a breaking Markov chain with the state space $\{0,1,\dots,n+1\}$ and the initial distribution $p = \{p_{0k}\}$.

Consider the condition $p_{l,l+k} = 0$ for $k > 1,\ \ p_{n+10} = 1$. Let $1 \leq l \leq n$. Since $\{\xi_k\}$ must first hit 1 in order to hit $n+1$, then

$$P\{\bar{\omega}_{l,n+1} \geq 1\} = P\{\bar{\omega}_{0,n+1} \geq 1\} \,/\, P\{\bar{\omega}_{0l} \geq 1\}.$$

By definition, we have $\gamma_l = P\{\bar{\omega}_{0l} \geq 1\} = P\{\bar{\nu}_l \geq 1\}$.

The quantity $\bar{\nu}_j$ can be represented in the form

$$\bar{\nu}_j = \bar{\gamma}_j(1+\bar{\zeta}_j),$$

where $\bar{\gamma}_j = \min\{1,\ \bar{\nu}_j\}$; $\ \xi_j = \max\{0,\ \bar{\nu}_j - 1\}$. Here $\bar{\gamma}_j$ characterizes presence of hits into state j ($\bar{\gamma}_j = 1$) or lack ($\bar{\gamma}_j = 0$) ; $\ 1+\bar{\zeta}_j$ is the number of hits into the state j "from below" under the supposition that one hit has occurred (evidently, the first hit always occurs "from below"). Note that $\bar{\zeta}_j$ equals the number of transitions from j into $j-1$, then into j, and has a geometric distribution with the parameter α_j because of the Markov chain property

$$P\{\bar{\zeta}_j = k\} = (1-\alpha_j)\alpha_j^k,\ \ \zeta_j = E\bar{\zeta}_j = \alpha_j/(1-\alpha_j).$$

Random variables $\bar{\zeta}_j$ and $\bar{\gamma}_j$ are mutually independent, therefore

$$\nu_j = E\bar{\nu}_j = E\bar{\gamma}_j(1+E\bar{\zeta}_j) = \gamma_j/(1-\alpha_j),$$
$$\mathcal{D}\bar{\nu}_j = E\bar{\gamma}_j^2\, E(1+\bar{\zeta}_j)^2 - (\gamma_j/(1-\alpha_j))^2 = \gamma_j(1+\alpha_j-\gamma_j)/(1-\alpha_j)^2.$$

Denote the number of hitting states i before hitting 0 or $n+1$ "from above"by $\bar{u}_i$. We have

$$\bar{u}_i = \sum_{j>i} \bar{\omega}_{0j}\eta_{ji}, \quad u_i = \bar{E}u_i = \sum_{j>i} \omega_{0j}p_{ji},$$

$$\omega_{0i} = (\nu_i + u_i)/(1-p_{ii}), \quad \omega_{0i}p_{ii} - \omega_{0i} = \nu_i + u_i.$$

In the last-but-one formula, we made use of the fact that the number of successive transitions from i to i is a random variable with a geometric distribution with the parameter p_{ii}. Taking into account that the chain path is stopped when hitting 0 or $n+1$, the formula for $\mathcal{D}\tilde{\tilde{\omega}}_{0,n+1}$ from Theorem 3.1 is transformed as follows

$$\mathcal{D}\tilde{\tilde{\omega}}_{0,n+1} = \sum_{k=0}^{n} \beta_k^{-1} \left\{ \omega_{0k} \sum_{l=k}^{n+1} p_{kl}\, \omega_{l,n+1}^2 + \sum_{l+k}^{n+1} \varphi_{0l}\, p_{kl}\, \omega_{k,n+1}^2 - \omega_{0k}^2 \omega_{k,n+1} \right\} +$$
$$+ \beta_{n+1}^{-1} \left\{ \omega_{0,n+1}[2\omega_{n+1,n+1} - 1 - \omega_{n+1,n+1}^2] \right\}.$$

Here $\omega_{n+1,n+1} = 1$. Hence, the addend with the multiplier β_{n+1}^{-1} disappears. Substituting the evident relation $\nu_{k+1} = \omega_{0k}p_{k,k+1}$ and the representations for ν_i, u_i, $\omega_{k,n+1}$ obtained above, in the formula, we have

$$\mathcal{D}\tilde{\tilde{\omega}}_{0,n+1} = \sum_{k=0}^{n} \beta_k^{-1} \left\{ \omega_{0k}p_{kk}\, \omega_{k,n+1}^2 + \omega_{0k}p_{kk}\, \omega_{k+1,n+1}^2 + u_k \omega_{k,n+1}^2 - \omega_{0k}\omega_{k,n+1}^2 \right\} =$$
$$= \sum_{k=0}^{n} \beta_k^{-1} \left\{ \nu_{k+1}\omega_{k+1,n+1}^2 + \omega_{k,n+1}^2 \left[\omega_{0k}p_{kk} - \omega_{0k} + u_k \right] \right\} =$$
$$= \gamma_{n+1}^2 \sum_{k=0}^{n} \beta_k^{-1} \left\{ \frac{\nu_{k+1}}{\gamma_{k+1}^2} - \frac{\nu_k}{\gamma_k^2} \right\} =$$
$$= \gamma_{n+1}^2 \sum_{k=0}^{n} \beta_k^{-1} \left\{ \gamma_{k+1}^{-1}(1-\alpha_{k+1})^{-1} - \gamma_k^{-1}(1-\alpha_k)^{-1} \right\}.$$

The test for the equality $E\tilde{\tilde{\gamma}}_{n+1} = E\bar{\gamma}_{n+1}$ is similar to that layed out in Theorem 2.1. Since $\tilde{\tilde{\omega}}_{0,n+1} = \tilde{\tilde{\gamma}}_{n+1}$, the theorem is proved.

11.1 Splitting and labelling technique.

Let the condition $p_{l,l+k} = 0$, $k > 1$, $p_{n+10} = 1$ be valid. Compare the particle (i,j) $i,j = 1,2,\ldots,s+1$ to the chain ξ_n path beginning in the state j $(\xi_0 = j)$ and being in the state i. Determine the branching random process $\{Y_k = \left(y_k^{1,1},\ldots,y_k^{n+1,n+1}\right)\}$ where $y_k^{i,j}$ is the number of particles of the type (i,j). Let $y_1^{1,1}$ be a random variable taking the value β_1 with probability p_{01} and the value 0 with probability $1-p_{01} = p_{00}$; $y_1^{i,j} = 0$, $i = 2,\ldots,n+1$. At every step, a particle of the type (i,j) (where $i,j = 1,\ldots,n+1$),

$j > i$ is turned into β_{i+1}/β_i particles of the type $(i+1,\, i+1)$ and one particle of the type $(i+1,\, j)$ with probability $p_{i,i+1}$; and into one particle of the type (l, j) with probability p_{il}, $l = i, i-1, \dots, j$; and is destroyed with probability $1 - \sum_{k=j}^{i} p_{ik} - p_{i,i+1}$. Note that particles of the type (i, j) with $j > i$ and $(n+1,\, n+1)$ do not appear in the process.

The process $\{Y_k\}$ describes the result of the application of the splitting and labelling technique to the simulation of a chain $\{\xi_k\}$ beginning in the state 0 and breaking when hitting 0 or $n+1$.

Denote by $\hat{\nu}_k$ the sum $\sum_{k=1}^{\infty}\sum_{j=1}^{n+1} y_k^{i,j}$, where only the particles, obtained from those with a smaller first index, i.e. caused by the chain jump "up", are taken into account.

Designate

$$\hat{\gamma}_{n+1} = \sum_{k=1}^{\infty}\sum_{j=1}^{n} y_k^{n+1,j}\quad \widehat{T} = \sum_{k=1}^{\infty}\sum_{i,j=1}^{n+1} y_k^{i,j} + 1,$$

$$\hat{\hat{\gamma}}_{n+1} = \hat{\gamma}_{n+1}/\beta_n,\quad \hat{\hat{\nu}}_j = \hat{\nu}_j/\beta_{j-1},$$

where $j = 1, 2, \dots, n+1$. Note that $\hat{\hat{\gamma}}_{n+1} = \hat{\hat{\nu}}_{n+1}$. Quantities $\hat{\hat{\nu}}_j$ are the unbiased estimators of quantities ν_j (equal to the expectation of number of hits into state j for the chain ξ_k beginning in the state 0 $(\xi_0 = 0)$ and breaking when hitting 0 or $n+1$). The following theorem states the fact and expressions for estimators variances.

Theorem 11.2. *If the conditions of Theorem 11.1 are fulfilled, then the process $\{Y_k\}$ is degenerated with probability* 1 *and*

$$E\hat{\hat{\nu}}_j = \nu_j,\quad \mathcal{D}\hat{\hat{\nu}}_j/\nu_j^2 = \sum_{k=0}^{j-1}\beta_k^{-1}C_k,\quad j = 2, \dots, n+1,$$

$$\hat{\hat{\nu}}_1 = \bar{\nu}_1,\quad C_k = \gamma_{k+1}^{-1}(1+\alpha_{k+1}) - \gamma_k^{-1}(1+\alpha_k) \geq 0,\quad E\widehat{T} = \sum_{i=0}^{n}\pi_i\beta_i/\pi_0.$$

Proof of Theorem 11.2. From the construction of process $\{Y_k\}$ it is clear that this process can be described in the following, equivalent way.

1. With probability p_{00}, all the quantities $\hat{\nu}_i$ are equal to 0, and the process $\{Y_k\}$ is degenerated: $Y_k = 0,\ k = 0, 1, \dots$.

2. With probability p_{01}, let us form β_1 independent chain $\{\xi_n\}$ paths with $\xi_0 = 1$ before hitting 0 or $n+1$ $\{\xi_n(\alpha)\}$, $\alpha = 1, 2, \dots, \beta_1$.

3. Assume

$$y_k^{j,1} = \sum_{\alpha=1}^{\beta_1}\varkappa(\xi_k(\alpha) = 1),$$

where $\varkappa(\xi_k(\alpha) = 1)$ is the corresponding event indicator.

4. Denote points of hitting state 2 from below by $n_{1,2}, \dots, n_{2,\hat{\nu}_2}$ (If $\hat{\nu}_2 = 0$, then $\hat{\nu}_i = 0$, $i > 2$ and the process begins again).

5. Let us form $(\beta_2/\beta_1 - 1)$ chains $\{\xi_k\}$ for every instant $n_{1,2}, \dots, n_{2,\hat{\nu}_2}$; for each of the chains with $\xi_0 = 2:\ \{\xi_k(\alpha, n_{2,j})\},\ \alpha = 1, \dots, \beta_2/\beta_2 - 1,\ j = 1, 2, \dots, \hat{\nu}_2$ the simulation process is broken when hitting 2 or $s+1$.

6. Set

$$y_s^{j,2} = \sum_{k=1}^{\hat{\nu}_2} \sum_{\alpha=1}^{\beta_1/\beta_2 - 1} \varkappa\left(\xi_{s+n_{2,k}}(\alpha, n_{2,k}) = j\right), \quad s = 1, 2, \dots$$

7. Processes 4–6 are repeated with state 2 for i substitution, β_i/β_{i-1} for β_2/β_1 for $i = 3, 4, \dots, n$.

Thus, the process $\{Y_k\}$ path can be represented as the sum of indicators of a random number of chain $\{\xi_k\}$ paths constructed before hitting $n+1$ or i and starting with i $(i = 1, 2, \dots, n)$.

Here

$$\hat{\nu}_1 = \bar{\nu}_1, \quad \hat{\nu}_{k+1} = \sum_{i=1}^{\hat{\nu}_k} \sum_{\alpha=1}^{\beta_k/\beta_{k-1}} \bar{\Delta}_k(i, \alpha), \quad k = 1, 2, \dots, n, \tag{11.1}$$

where $\bar{\Delta}_k(i,\alpha),\ i = 1, 2, \dots, \hat{\nu}_k,\ \alpha = 1, 2, \dots,\ \beta_k/\beta_{k-1}$ are independent versions of the random variable $\bar{\Delta}_k$ defined as the number of hittings into the state $k+1$ from k by the chain $\{\xi_n\}$ before hitting $n+1$ or $k-1$.

By setting $\beta_k \equiv 1$, we have

$$\bar{\nu}_{k+1} = \sum_{i=1}^{\bar{\nu}_k} \bar{\Delta}_k(i). \tag{11.2}$$

Since $\bar{\Delta}_k$ and $\bar{\nu}_k$ do not exceed $\bar{T}$, and $E\bar{T}$ and $\mathcal{D}\bar{T} < \infty$, then the expectation and the variance of $\bar{\nu}_k$ and $\bar{\Delta}_k$ are finite.

Whence, and from the recurrence formulas (11.1), the finiteness of the expectations and variances of $\hat{\nu}_i$ follows, as well as the degeneration of $\{Y_k\}$ with probability 1.

Computing the expectations of the left and right sides of (11.1) and (11.2), we have

$$\nu_{k+1} = \nu_k \Delta_k, \quad E\hat{\nu}_{k+1} = (\beta_k/\beta_{k-1})\, E\hat{\nu}_k \Delta_k,$$

where $\Delta_k = E\bar{\Delta}_k,\ \nu_k = E\bar{\nu}_k$. Hence, we have

$$\Delta_k = \nu_{k+1}/\nu_k, \quad E\hat{\nu}_1 = \nu_1, \quad E\hat{\nu}_{k+1} = \nu_{k+1}\beta_k,$$
$$E\hat{\hat{\nu}}_{k+1} = \nu_{k+1}.$$

Computing the expectations of the squares of the left and the right sides of (11.2), we have

$$\begin{aligned} \mathcal{D}\bar{\nu}_{k+1} &= \mathcal{D}\bar{\nu}_k \Delta_k^2 + \nu_k \mathcal{D}\bar{\Delta}_k, \\ \mathcal{D}\bar{\Delta}_k &= \left(\mathcal{D}\bar{\nu}_{k+1}/\nu_{k+1}^2 - \mathcal{D}\bar{\nu}_k/\nu_k^2\right) \frac{\nu_{k+1}^2}{\nu_k}. \end{aligned} \tag{11.3}$$

Compute now the expectations of the squares the left and right sides of (11.1)

$$\mathcal{D}\bar{\nu}_{k+1} = \mathcal{D}\hat{\nu}_k \Delta_k^2 \left(\frac{\beta_k}{\beta_{k+1}}\right)^2 + E\hat{\nu}_k \mathcal{D}\bar{\Delta}_k \frac{\beta_k}{\beta_{k-1}}.$$

Whence, and taking (11.3) into account, we have

$$\frac{\mathcal{D}\hat{\hat{\nu}}_{k+1}}{\nu_{k+1}^2} - \frac{\mathcal{D}\hat{\hat{\nu}}_k}{\nu_k^2} = \frac{1}{\beta_k}\left(\frac{\mathcal{D}\bar{\nu}_{k+1}}{\nu_{k+1}^2} - \frac{\mathcal{D}\bar{\nu}_k}{\nu_k^2}\right). \tag{11.4}$$

Designate $\bar{\nu}_0 = 1$, $\mathcal{D}\bar{\nu}_0 = \mathcal{D}\hat{\hat{\nu}}_0 = 0$. By summing the left and right sides of equality (11.4) for $k = 0, 1, \ldots, j-1$, we have

$$\mathcal{D}\hat{\hat{\nu}}_j / \nu_j^2 = \sum_{k=0}^{j-1} \beta_k^{-1} \left(\frac{\mathcal{D}\bar{\nu}_{k+1}}{\nu_{k+1}^2} - \frac{\mathcal{D}\bar{\nu}_k}{\nu_k^2}\right).$$

Substituting the expression for $\mathcal{D}\bar{\nu}_k$ and ν_k obtained from the proof of Theorem 11.1 in this relation, we will find the formulas for variances indicated in the conclusion of Theorem 11.2. The statement $C_k \geq 0$ follows from (11.3).

Formulas $E\hat{\nu}_k = \beta_{k-1}\nu_k$ and $E\hat{\hat{\nu}}_k = \nu_k$ were obtained above. To compute $E\widehat{T}_k$, it should be noted that $\widehat{T}$ can be represented as

$$\widehat{T} = \sum_{i=1}^{n} \sum_{\alpha=1}^{a_i} [\bar{\rho}_i(\alpha) + 1] + 1,$$

where

$$a_i = (\beta_i/\beta_{i-1})\hat{\nu}_i + \hat{u}_i, \quad \hat{u}_i = \sum_{k=1}^{\infty} \sum_{j=i}^{n} y_k^{i,j}.$$

Summation is taken over all the particles generated by those with a larger first index, i.e. particles caused by $\{\xi_k\}$ jumps "down"; $\bar{\rho}_i(\alpha)$ are independent versions of the random variable $\bar{\rho}_i$ equal to the number of chain $\{\xi_k\}$ hittings in succession from i into i. As was mentioned above, $\bar{\rho}_i$ has a geometric distribution with the parameter p_{ii}. Morever,

$$\hat{u}_i = \sum_{j=1}^{i} \sum_{\alpha=1}^{b_i} \bar{\tau}_{ij}(\alpha),$$

where $b_i = \hat{\nu}_j(\beta_j/\beta_{j-1} - 1)$, $j \geq 2$, $b_1 = \hat{\nu}_1\beta_1$, $\bar{\tau}_{ij}(\alpha)$ are independent versions of the random variables $\bar{\tau}_{ij}$; $\bar{\tau}_{ij}$ equals the number of hits into states i up by the chain $\{\xi_k\}$ simulated from $\xi_0 = i$ before hitting j (or $n+1$).

Further, the mean number of hittings i up by the chain simulated from $j \le i$ before returning to j is equal to $E\bar{u}_i/E\bar{\nu}_j$. Therefore,

$$E\bar{\tau}_{ij} = E\bar{u}_j/E\bar{\nu}_j,$$

$$E\hat{u}_i = \sum_{j=1}^{i} E\hat{\nu}_j(\beta_j/\beta_{j-1} - 1)\, E\bar{u}_i/E\bar{\nu}_j + E\hat{\nu}_1\beta_1 E\bar{u}_i/E\bar{\nu}_1 =$$

$$= E\bar{u}_i \left\{ \sum_{j=2}^{i} (\beta_j/\beta_{j-1}) + \beta_1 \right\} = \beta_i u_i.$$

Now we have

$$E\widehat{T} = 1 + \sum_{i=1}^{n} (1 + E\bar{\rho}_i)\beta_i(u_i + \nu_i) = 1 + \sum_{i=1}^{n} \beta_i(u_i + \nu_i)/(1 - p_{ii}) =$$

$$= 1 + \sum_{i=1}^{n} \omega_{0i}\beta_i = \sum_{i=0}^{n} \pi_i\beta_i/\pi_0.$$

Theorem 11.2 is complete.

11.2. Techniques comparison.
We can choose the quantity

$$E\bar{T}\mathcal{D}\bar{\gamma}_{n+1}/ET_{(b)}\mathcal{D}\gamma_{n+1(b)},$$

as the efficiency criterion where $T_{(b)}$ is the lifetime of all the particles generated by a branching process and $\mathcal{D}\gamma_{n+1(b)}$ is the variance of the quantity γ_{n+1} estimator (hitting $n+1$ before returning to 0 probability). We have a roulette technique $\{Z_n\} : T_{(b)} = \tilde{T}$, $\gamma_{n+1} = \tilde{\tilde{\gamma}}_{n+1}$ and the labelling technique with the process $\{Y_n\} : T_{(b)} = \widehat{T}$, $\gamma_{n+1}(b) = \hat{\hat{\gamma}}_{n+1}$. By virtue of Theorems 11.1. and 11.2,

$$E\widehat{T} = E\tilde{T} = \sum_{j=0}^{s} \pi_j\beta_j/\pi_0,$$

and it is sufficient to compare $\mathcal{D}\tilde{\tilde{\gamma}}_{n+1}$ and $\mathcal{D}\hat{\hat{\gamma}}_{n+1}$.

Theorem 11.3. *Under the assumptions of Theorem 11.1,*

$$\mathcal{D}\tilde{\tilde{\gamma}}_{n+1} \ge \mathcal{D}\hat{\hat{\gamma}}_{n+1},$$

where equality holds if and only if $p_{k0} = 1 - p_{k,k+1}$ $(k = 1, 2, \ldots, n)$ *or* $\beta_k \equiv 1$.

Proof. Using the formulas of Theorems 11.1 and 11.2, we have

$$\mathcal{D}\tilde{\tilde{\gamma}}_{n+1} - \mathcal{D}\hat{\hat{\gamma}}_{n+1} = \gamma_{n+1}^2 \sum_{k=1}^{n} \left(\beta_{k-1}^{-1} - \beta_k^{-1}\right) \gamma_k^{-1} \{(1-\alpha_k)^{-1} - (1+\alpha_k)\}. \tag{11.5}$$

Each of the right side addends is ≥ 0. Equality is attained only in two cases: 1) $\beta_k \equiv 1$ and 2) $\alpha_k \equiv 0$. For $\alpha_k \equiv 0$, any state is hit no more than once, whence $p_{k0} = 1 - p_{k,k+1}$, $(k = 1, 2, \dots, n)$. Thus, the theorem is complete.

Theorem 11.3 makes it possible to draw a conclusion that the labelling method is more efficient than the roulette method but, for small values of α_k, there will be little difference in the efficiency determined by formula (11.5).

11.4 Optimal experimental design.

The optimal choice of the coefficient β_k has a direct analogy with the problems of the mathematical theory of experimental design developed for natural tests (see Section 5). As in this theory, reasonable suggestions for choosing the coefficients β_i (plays an experimental design role) can be obtained by abandoning the integrity of β_i/β_{i-1}.

Definition. For the problem under consideration a collection of numbers $1 = \beta_0 \leq \beta_1 \leq \dots \leq \beta_{n+1} < \infty$ used in branching methods is called the branching simulation design (BSD).

Taking the results of Theorem 11.3 into account, we will further confine ourselves to the optimal designs of the labelling technique. Note that analogous approaches and results are valid for the roulette technique.

Let the parameters of the form ν_{n+1}, $\sum_{i=1}^{n+1} \theta_i \nu_i$ estimator be investigated, where θ_i are given coefficients.

Introduce the following optimum criteria of BSD

$$\mathcal{D}\hat{\nu}_{n+1} E\widehat{T} \to \min_{\beta}, \tag{*}$$

$$\left\{\det \left\| Cov\left(\hat{\nu}_k, \hat{\nu}_l\right)\right\|\right\}^{1/(n+1)} E\widehat{T} \to \min_{\beta}. \tag{**}$$

The probabilistic sense of the first criterion is evident, and the sense of the second is known from experimental design theory [31].

From the proof of Theorem 11.2, it follows that the criterion (*) is independent of β_{n+1}. As is further seen in Theorem 11.4 a similar fact is valid for (**). Therefore, from now on, by BSD a vector $\beta = (\beta_0, \dots, \beta_n)$ is meant, where $\beta_0 = 1$ and $\beta_i \geq \beta_{i-1}$.

Designate

$$C_k = \gamma_{k+1}^{-1}(1 + \alpha_{k+1}) - \gamma_k^{-1}(1 + \alpha_k), \quad k = 0, 1, \dots, n.$$

The optimal BSD for criteria (*) and (**) is given in explicit form by the following theorem:

Theorem 11.4. *Under the conditions of Theorem 11.2, the following conclusions are valid for the labelling method.*

1. *If the sequence $\{C_k/\pi_k\}$ is not decreasing, then by the criterion (*) the quantities $\beta_k = \beta_k^* = (C_k\pi_0/C_0\pi_k)^{1/2}$ form an optimal BSD, which is unique.*

2. If the sequence $\{\pi_k\}$ is not increasing then the by the criterion (**) the quantities $\beta_k = \beta_k^* = \pi_0/\pi_k$ form an optimal BSD, which is unique.

Proof of Theorem 11.4.

1. By virtue of Theorem 11.2, $\mathcal{D}\hat{\hat{\nu}}_{n+1}E\widehat{T}$ has the form

$$\sum_{k=0}^{n} \beta_k^{-1} C_k \sum_{k=0}^{n} \beta_k \pi_k/\pi_0.$$

By Schwartz' inequality, this value is either larger than or equal to

$$\left[\sum_{k=0}^{n} (C_k/\pi_k)^{1/2}\right]^2 \Big/ \pi_0$$

and equality is attained for $\beta_0 = 1$ if and only if

$$\beta_k = \beta_k^* = [C_k\pi_0/C_0\pi_k]^{1/2}.$$

Since the monotonicity of the right side with respect to k is supposed, then $1 = \beta_0^* \leq \beta_1^* \leq \ldots \beta_n^* < \infty$, i.e. numbers β_k^* make up a BSD.

2. For computations of Cov $\left(\hat{\hat{\nu}}_k, \hat{\hat{\nu}}_l\right)$ note that, by virtue of equality (11.1), the distribution of the quantity $\hat{\hat{\nu}}_{k+1}/\hat{\hat{\nu}}_k$ does not depend on $\hat{\hat{\nu}}_k, \ldots, \hat{\hat{\nu}}_1$. Here

$$E\left(\hat{\hat{\nu}}_{k+1}/\hat{\hat{\nu}}_k\right) = \nu_{k+1}/\nu_k.$$

Consequently,

$$E\hat{\hat{\nu}}_{k+l}\hat{\hat{\nu}}_k = E\hat{\hat{\nu}}_k^2 E\frac{\hat{\hat{\nu}}_{k+1}}{\hat{\hat{\nu}}_k}\ldots E\frac{\hat{\hat{\nu}}_{k+l}}{\hat{\hat{\nu}}_{k+l-1}} = \mathcal{D}\hat{\hat{\nu}}_k\frac{\nu_{k+l}}{\nu_k} + \nu_k\nu_{k+l},$$

$$Cov\left(\hat{\hat{\nu}}_k, \hat{\hat{\nu}}_{k+l}\right) = \left(\mathcal{D}\hat{\hat{\nu}}_k\right)\nu_{k+l}/\nu_k, \quad l = 1, 2, \ldots, n+1-k,$$

where $\mathcal{D}\hat{\hat{\nu}}_k$ can be derived from the formulas in Theorem 11.2. We have

$$\det\left\|\text{Cov}\left(\hat{\hat{\nu}}_k, \hat{\hat{\nu}}_j\right)\right\|_1^{n+1} = \nu_1^2\ldots\nu_{n+1}^2 \det\Phi(\beta, C),$$

where $\Phi(\beta, C)$ is a matrix in the form of

$$\begin{Vmatrix} a_1 & a_1 & \ldots & a_1 \\ a_1 & a_1+a_2 & \ldots & a_1+a_2 \\ \vdots & \vdots & \ldots & \vdots \\ a_1 & a_1+a_2 & \ldots & a_1+\ldots+a_{n+1} \end{Vmatrix}$$

and $a_i = \beta_{i+1}^{-1} C_{i+1}$. Subtracting the first line from the others, then the second from the third, etc, then the third from the fourth, etc., we have

$$\det \Phi(\beta, C) = a_1 \dots a_{n+1} = \beta_1^{-1} \dots \beta_n^{-1} C_0 \dots C_n.$$

Now from the inequality between the arithmetic mean and the geometric mean, we have

$$E\widehat{T}\left\{\det \left\| Cov\left(\hat{\nu}_k, \hat{\nu}_j\right)\right\|\right\}^{1/(n+1)} =$$

$$= \left[\nu_1^2 \dots \nu_{n+1}^2 \beta_1^{-1} \dots \beta_n^{-1} C_0 \dots C_n\right]^{1/(n+1)} \sum_{k=0}^{n} \beta_k \pi_k / \pi_0 \geq$$

$$\geq \left[\nu_1^2 \dots \nu_{n+1}^2 C_0 \dots C_n \prod_1^n \pi_k\right]^{1/(n+1)} (n+1)/\pi_0^{n/(n+1)}$$

and equality is attained if and only if $\beta_i = \beta_i^* = \pi_i/\pi_0$ and $1 = \beta_0^* \leq \beta_1^* \leq \dots \leq \beta_n^*$ by the monotonicity of π_i. Q.E.D.

The formulas for the BSD which is optimal by criterion (**), have a simple form and are convenient to use.

11.5 The technique asymptotic efficiency evaluating.

Let us evaluate the efficiency of the technique under the optimal choice of BSD by way of an example of the Markov chain with states equal to the number of requests in the queueing system of the type $M/M/1$ at the arrival of the requests and the service completion instances. Let the system load be $\rho = \lambda/\mu$, where λ is the entering flow and $1/\mu$ is the mean service time intensity. It is known [20] that, in this case, the matrix P elements have the form

$$p_{01} = \rho, \; p_{00} = 1 - \rho, \; p_{k,k+1} = 1 - p_{k,k-1} = \lambda/(\lambda + \mu) = \rho/(1+\rho),$$
$$\pi_0 = 1 - \rho, \quad \pi_k = (1-\rho)\rho^k, \quad k \geq 1.$$

Let us suppose that the system returns to 0 with probability 1 when the number of requests achieves $n+1$.

For $\rho \to 0$, direct calculations give $\alpha_k \to 0, \;\; \gamma_k \sim \rho^k, \;\; \pi_k \sim \rho^k$. Whence, by virtue of Theorems 11.2 and 11.4,

$$\mathcal{D}\hat{\nu}_{n+1} \sim \sum_{k=0}^{n} \beta_k^{-1} \rho^{-(k+1)}, \quad E\widehat{T} \sim \sum_{k=0}^{n} \beta_k \rho^k, \quad \beta_k^* \sim \rho^{-k}.$$

We obtain the asymptotic efficiency value

$$E\bar{T}\mathcal{D}\bar{\nu}_{n+1} \,/\, E\widehat{T}\mathcal{D}\hat{\nu}_{n+1} \sim \rho^{-n}/(n+1)^2.$$

For example, for $n = 5, \;\; \rho = 1/10$ the efficiency value equals 227, for $s = 10, \;\; \rho = 1/3$, the efficiency value is 521.

Thus, for small load values, and, particularly, in this case, direct simulation is labor-consuming and the gain in time (under fixed accuracy) is significant.

CHAPTER 4. GENERAL MARKOV CHAINS

In this chapter, we will extend the main results of Chapters 2 and 3 to the general Markov chain case. As before, we will consider two classes of chains: absorbing and ergodic. The results of Chapter 2 concerning the transition behaviour (Section 2.2) and absorbing chains (Section 2.3) are almost completely extended to the case of general Markov chains. As for ergodic chains, applying the regeneration technique has the following essential peculiarity. For the finite-dimensional case, the role of the regeneration state can be played by any state of the chain, and for the case of general chains, this role is played by some functions defined on the state space. Finding this function may be an intricate problem, since its general solution technique not yet developed. In this connection, the case of a queueing system has been investigated by I.N. Kovalenko [67]. Some interesting examples can be also found in work [98].

1. MAIN DEFINITIONS AND NOTATION

The detailed development of the concepts and results considered in this section can be found in the monograph [89].

A sequence of random variables $X_0, X_1, \dots, X_n, \dots$ is called a Markov chain if

$$P\{X_{n+1} < x \mid X_n < x_n, \dots, X_0 < x_0\} = P\{X_{n+1} < x \mid X_n < x_n\}.$$

A Markov chain is defined by the initial distribution $\lambda(dx) = P\{X_0 \in dx\}$ and probability of transition in one step $P_{(l)}(x, dy) = P\{X_{l+1} \in dy \mid X_l = x\}$. Further, we will consider only homogeneous chains for which $P_{(l)}(x, dy) \equiv P(x, dy)$, $l = 1, 2, \dots$. These will be noted by (λ, P). Note that all the results relating to the transition behaviour, i.e., to the chain behavior in a given finite number of steps, are automatically extended to the case of nonhomogeneous chains.

We will consider that the random variables $X_0, X_1 \dots$ defined on the same probabilistic space. Let us denote by $\mathcal{X}$ the set of possible values of these variables and by $\mathcal{A}$ a σ-algebra of its subsets. Consider the σ-algebra $\mathcal{A}$ to be countably generated. Let $P(x, dy)$ be a measurable function of the first argument under a fixed second argument and a measure on $(\mathcal{X}, \mathcal{A})$ under the fixed first argument, where $\int_{\mathcal{X}} P(x, dy) \le 1$. Further, $\int_{\mathcal{X}}$ will be denoted by $\int$.

Designate $g(x) = 1 - \int P(x, dy)$. In the case of $g(x) \not\equiv 0$, complement the set $\mathcal{X}$ by an element Δ and define the chain $\{X'_n\,;\, n \ge 0\}$ for any chain $\{X_n\,;\, n \ge 0\}$ with the transition probability $P(x, dy)$ by the following relations

$$P'(x, dy) = P(x, dy) \quad \text{for} \quad x, y \in \mathcal{X},$$
$$P'(x, \Delta) = g(x), \qquad P'(\Delta, \Delta) = 1.$$

Thus, if for some n $X'_n = \Delta$, then $X'_{n+1} = \Delta$, $X'_{n+2} = \Delta, \dots$. Further, the prime will be omitted in the convention that the Markov chain is determined in the way indicated above for the case of $g(x) \not\equiv 0$.

Consider absorbing and ergodic chains.

Definition 1.1. A Markov chain (λ, P) is called absorbing if, with probability 1, there exists such a number n that $X_k = \Delta$ for $k \geq n$.

A sequence of values of the random variables $X_0, X_1, \ldots$ (these values and the random variables themselves have a similar notation, which should not lead to confusion) is called the Markov chain path.

The quantity $\eta = f(X_0, X_1, \ldots)$ where f is a measurable function on the space $\mathcal{X} \otimes \mathcal{X} \otimes \ldots = \mathcal{X}^\infty$ is called the functional of the chain.

The quantity

$$J_N = \sum_{k=0}^{N} h(X_k),$$

where h is a measurable function and N is a fixed integer, is called the linear functional of the path (in the transition regime). Denote a limit $\lim_{N \to \infty} J_N$, if such a limit exists, by J.

Let us introduce several additional concepts for the definition of the ergodic Markov chains. Define the function $h_B^\infty(x)$ in the following way

$$h_B^\infty(x) = P\{\, X_n \in B \text{ infinitely often } \mid X_0 = x \,\},$$

where the statement "$X_n \in B$ infinitely often"is equivalent to

$$\bigcap_{m=0}^{\infty} \bigcup_{n=m}^{\infty} \{X_n \in B\} = \varnothing.$$

We will speak that a Markov chain is recursive in the Harris sense if a nonnegative measure μ on $(\mathcal{X}, \mathcal{A})$, such that $\mu(\mathcal{X}) \geq 0$ and

$$h_B^\infty(x) \equiv 1 \text{ for } x \in \mathcal{X} \text{ and } B \in \{B \in A;\ \mu(B) \geq 0\},$$

exists. It is known that for Markov chains which are recursive in the Harris sense, there exists a unique up to a constant multiplier measure π (nonnegative), such that

$$\pi(dy) = \int P(x, dy)\, \pi(dy). \tag{1.1}$$

If $\pi(\mathcal{X}) < \infty$, then chain is called positively recursive. In this case, the measure π is considered to be normalized: $\pi(\mathcal{X}) = 1$, i.e., to be a probability measure. Such a measure satisfying relation (1.1) is called a stationary measure (stationary distribution).

We state that a set B is accessible from x if there exists such n that $P^n(x, B) > 0$ (denote this fact by $x \to B$).

A chain is called irreducible as well as μ irreducible if $x \to B$ for any $x \in \mathcal{X}$ and any μ positive set B (i.e. such B that $\mu(B) > 0$) for some nonnegative measure μ such that $\mu(\mathcal{X}) > 0$.

The following statement is valid for any irreducible Markov chains.

Statement 1.1. *If a Markov chain (λ, P) is μ irreducible, then there exist a μ positive function $s(x)$ (i.e. $\int_B s(x)\,\mu(dx) > 0$ for $\mu(B) > 0$) and a probability measure ν such that for some $m_0 \geq 1$,*

$$P^{m_0}(x, B) \geq s(x)\,\nu(B),$$

for any $x \in \mathcal{X}, \;\; B \in \mathcal{A}$.

By the definition given above, a Markov chain which is recursive in the Harris sense is irreducible.

A sequence $\mathcal{X}_0, \mathcal{X}_1, \dots, \mathcal{X}_{m-1}$ consisting of nonempty nonintersecting sets is called an m cycle if, for any $i = 0, 1, \dots, m-1$ and for any $x \in \mathcal{X}_i$, the condition

$$P(x,\ \mathcal{X} \backslash \mathcal{X}_j) = 0 \;\text{ for }\; j = i + 1 (\text{mod } m).$$

is valid. It is known that each irreducible chain for some fixed d has a d cycle, such that for any d' cycle, d'/d is an integer.

If $d = 1$, then the chain is called aperiodic.

A chain is called positively recursive in the Harris sense if it is simultaneously positively recursive and recursive in the Harris sense.

Definition 1.2. A Markov chain is called ergodic if it is positively recursive in the Harris sense and aperiodic.

The following statement has the key significance.

Theorem 1.1. *A Markov chain is ergodic if and only if for any measure ν*

$$\lim_{n \to \infty} \|\nu P^n - \pi\| = 0,$$

where $\|\mu\| = \sup_{B \in \mathcal{A}} \mu(B) - \inf_{B \in \mathcal{A}} \mu(B)$.

From this theorem, it follows that for any integrable with respect to measure π function h and for any measure ν,

$$\nu P^n h - (h, \pi) = \int \nu(dx)\, P^n(x, dy)\, h(y) - \int \pi(dx)\, h(x) \to 0.$$

We can now evaluate a linear functional of stationary distribution as

$$J = (\pi, h) = \lim_{n \to \infty} \nu P^n h,$$

where h is an arbitrary integrable with respect to measure π function.

In the following sections, we will consider the estimators of the functionals J_N, J, J_{st} along the Markov chain path.

2. Functionals J_N and J estimators

Note that the statements and proofs of the theorems from Section 2 in Chapter 2 hold for general Markov chains if (h,π) is interpreted as $\int h(x)\,\pi(dx)$, Ph as $\int P(x,dy)\,h(y)$, $\sum_{y\in\mathcal{X}} P(x,y)h(y)$ is replaced by $\int P(x,dy)\,h(y)$, and the definition

$$\zeta_k(x,dy) = \begin{cases} 1, & \text{if } X_{k-1} = x, \quad X_k \in dy, \\ 0, & \text{otherwise}, \end{cases}$$

is introduced instead of $\zeta_k(x,y)$.

Therefore, the following theorem is valid.

Theorem 2.1. *Taking into account the indicated changes for the arbitrary (not necessary homogeneous) Markov chain introduced in Section 1 the theorems from Section 2 of Chapter 2 are valid.*

Consider absorbing Markov chains. Let $\tau+1$ be an absorption instant, i.e., $X_{\tau+1} = \Delta$. We will consider that $\sup_{x\in\mathcal{X}} h(x) < \infty$, $h(\Delta) = 0$. Suppose that $E\tau < \infty$. By virtue of the generalization of Wald's lemma (see Lemma 5.2 from monograph [89]), we have

$$E\sum_{n=0}^{\tau} h(X_n) \le E\,h(X_0) + ME\tau,$$

where $M = \sup_{n\ge 1} \{\text{ess sup } E\{h(X_n) \mid X_{n-1}\}\}$.

Since

$$E\,h(X_0) = \int \nu(dx)\,h(dx) \le \sup_{x\in\mathcal{X}} h(x) < \infty,$$

$$M \le \sup_{x\in\mathcal{X}} \int P(x,dy)\,h(y) \le \sup_{x\in\mathcal{X}} h(x) < \infty,$$

then $E\sum_{n=0}^{\tau} h(X_n) < \infty$.

Note that

$$E\sum_{0}^{\infty} h(X_n') = E\sum_{0}^{\tau} h(X_n) < \infty,$$

$$\lim_{N\to\infty} E\,J_{1(N)} = \sum_{n=0}^{\infty} \int P^n(x,dy)\,h(y)\,\nu(dx) = E\sum_{0}^{\tau} h(X_n) < \infty.$$

Therefore, the sequence

$$\sum_{n=0}^{\infty} \int P^n(x,dy)\,h(y)\,\nu(dx)$$

converges. Under the condition that this sequence converges, the results achieved in Section 3 of Chapter 2 have been proved for the case of general absorbing Markov chains in the monograph [29], and they can be also obtained by a technique similar to the finite-dimensional case.

Let $\widehat{J}_1$ and $\widehat{J}_2$ be the estimators defined in Section 2.3, $a_{ij} \equiv 1$.

The following statement is valid.

Theorem 2.2. *For an arbitrary homogeneous absorbing Markov chain* (ν, P) *and an arbitrary measurable with respect to* $(\mathcal{X}, \mathcal{A})$ *function* $h(x)$, *such that* $\sup_{x\in\mathcal{X}} h(x) < \infty$ *on condition* $E\tau < \infty$, $\widehat{J}_1$ *and* $\widehat{J}_2$ *are unbiased estimators of* J *and*

$$E\,\widehat{J}_1^2 = (\Psi,\ 2\varphi h - h^2), \quad E\,\widehat{J}_2^2 = (\Psi,\ h^2/g).$$

Here φ *is the iterative solution of the equation* $\varphi = P\varphi + h$, $\varphi = \lim_{N\to\infty}\sum_{k=0}^{N} P^k h$, *and* Ψ *is the iterative solution of the equation* $\Psi = \Psi P + \nu$, *where* $\Psi\, P_{(\,.\,)} = \int \Psi(dx)\, P(x, .)$.

Results for fictitious chain are produced similary.

3. Linear functional of the stationary distribution estimator

Let $\{X_n;\ n \geq 0\}$ be an ergodic Markov chain with a transition probability function in one step $P(x, dy)$ and a stationary distribution π. Let $h(x)$ be a measurable function on $(\mathcal{X}, \mathcal{A})$, $\tilde{h}(x) = h(x) - J_{st}$.

Assume that

$$\int |\tilde{h}(x)|\, \pi(dx) < \infty,$$

$$\int |\tilde{h}(x)|\, E\left\{\sum_{n=1}^{\tau(B)} h(X_n) \mid X_0 = x\right\} \pi(dx) < \infty \tag{3.1}$$

for any $B \in \mathcal{A}$, where $\tau(B) = \min\{n;\ X_n \in B\}$.

As was mentioned above (Statement 1.1), there exists an integer m_0, a nonnegative function $s(x)$, and a probabilistical measure ν such that

$$P^{m_0}(x, B) \geq s(x)\, \nu(B) \tag{3.2}$$

for $x \in \mathcal{X}$, $B \in \mathcal{A}$.

Set

$$G_{m_0,s,\nu} = \sum_{n=0}^{\infty} (P^{m_0} - s \otimes \nu)^n (1 + P + \ldots + P^{m_0 - 1}).$$

Theorem 3.1. *Under the conditions formulated above, the distribution of the quantity* $\sqrt{N}\,(\tilde{J}_N - J_{st})$ *converges to normal with zero expectation and* σ^2 *variance where*

$$\sigma^2 = (\pi, h^2) + 2m_0^{-1}\left[\sum_{m=1}^{m_0} (m_0 - m)(\pi,\ P^m \tilde{h}) + \sum_{m=1}^{m_0} (\pi,\ P^m G_{m_0,s,\nu}\, \tilde{h})\right].$$

This theorem is an analogue of Theorem 2.4.1. Its proof can be found in the monograph [89, Theorem 7.6], and it is based on the regeneration method which can be used for constructing estimators like in the finite chains case.

Further, for simplification of notation, take $m_0 = 1$ (the general case is considered similarly). Then

$$\sigma^2 = (\pi, \tilde{h}^2) + 2(\pi,\ PG_{s,\nu}\tilde{h}) = (\pi,\ 2\varphi\tilde{h} - \tilde{h}^2),$$

where $G_{s,\nu} = \sum_{n=0}^{\infty}(P - s\otimes\nu)^n$, φ is an iterative solution of the equation $\varphi = (P - s\otimes\nu)\varphi + \tilde{h}$.

The quantity $\sigma^2 = (\pi, 2\varphi\tilde{h} - \tilde{h}^2)$ is independent of ν and s as it follows from Theorem 3.1.

The following scheme is convenient for the investigation of estimators of the regeneration method. Let $\{Z_n\,;\ n \ge 0\}$ be a random process with the values $\{0,1\}$ and $Z_n = 0$ with probability $1 - s(X_n)$.

Let $X_0 \sim \nu(dx)$, $\alpha_0 = 0$, $\alpha_1 = \min\{n \ge 0,\ Z_n = 1\}$, $\alpha_2 = \min\{n > \alpha_1,\ Z_n = 1\},\ldots$ It is known [89] that α_i is a random variable and that the X_{α_i} distribution coincides with ν. By virtue of the Markov property $\{X_n\}$, we have that $Y(1), Y(2),\ldots, Y(k),\ldots$ and $b(1), b(2),\ldots, b(k),\ldots$ are independent and equally distributed random variables (IEDR), and the pairs $(Y(1),\ b(1))$, $(Y(2),\ b(2))\,,\ldots$ are also independent and equally distributed, where

$$Y(k) = \sum_{i=\alpha_{k-1}}^{\alpha_k - 1}\tilde{h}(X_i), \quad b(k) = \alpha_k - \alpha_{k-1}.$$

Consider the estimator

$$\tilde{J}_{3(k)} = \frac{1}{k}\sum_{i=1}^{k}Y(i)\Big/\left(\frac{1}{k}\sum_{i=1}^{k}b(i)\right).$$

Let us investigate the properties of the estimator $\tilde{J}_{3(k)}$ by means of a pair of conjugate equations similar to equations (3.1) and (3.2) from Chapter 3.

Set

$$\bar{\varphi}(x) = \left\{\sum_{i=0}^{\alpha_1}h(X_i)\ |\ X_0 = x\right\},$$

$$\bar{\varphi}^{(n)}(x) = \left\{\sum_{i=0}^{n_1}h(X_i)\ |\ x_0 = x\right\},$$

where $\alpha_1 = \min\{k \ge 0,\ Z_k = 1\}$, $n_1 = \min\{n, \alpha_1\}$. Since $\bar{\varphi}(x)$, $\bar{\varphi}^{(n)}(x) \ge 0$, then expectations (finite or infinite) of these quantities exist. Designate $\varphi(x) = E\bar{\varphi}(x)$, $\varphi^{(n)}(x) = E\bar{\varphi}^{(n)}(x)$.

Since $\bar{\varphi}(x) = \{Y(1)\ |\ X_0 = x\}$, then

$$EY(1) = \int\varphi(x)\,\nu(dx), \quad EY^2(1) = \int\bar{\varphi}^2(x)\,\nu(dx).$$

Evidently, the quantities $\bar{\varphi}^{(n)}(x)$ satisfy the relations

$$\bar{\varphi}^{(n+1)}(x) = \int \zeta(x, dy)\, \bar{\varphi}^{(n)}(x) + h(x), \tag{3.3}$$

where $\zeta(x, dy)$ is defined under the condition $X_0 = x$ and

$$\zeta(x, dy) \doteq \begin{cases} 1, & \text{if } X_1 \in dy, \quad Z_0 = 0, \\ 0, & \text{if } Z_0 = 1. \end{cases}$$

Since $P\{X_1 \in dy \mid Z_0 = 1, \; X_0 = x\} = \nu(dy)$, then

$$\begin{aligned} P(x, dy) &= P\{X_1 \in dy \mid X_0 = x\} = \\ &= P\{X_1 \in dy, \; Z_0 = 1 \mid X_0 = x\} + P\{X_1 \in dy, \; Z_0 = 0 \mid x_0 = x\} = \\ &= s(x)\nu(dy) + P\{X_1 \in dy, \; Z_0 = 0 \mid X_0 = x\}. \end{aligned}$$

Hence,

$$E\zeta(x, dy) = P\{X_1 \in dy, \; Z_0 = 0 \mid X_0 = x\} = P(x, dy) - s(x)\nu(dy).$$

Denote this value by $Q(x, dy)$. Calculating the expectations of both sides of relation (3.3), we obtain the equality

$$\varphi^{(n+1)}(x) = \int Q(x, dy)\, \varphi^{(n)}(y) + h(x), \quad n = 0, 1, \ldots$$

Moreover, $\varphi^{(0)}(.) = h(.) / \int \nu(dx)\, s(x)$.

Since $\varphi^{(n+1)}(x) \geq \varphi^{(n)}(x)$, then passing to the limit at $n \to \infty$, we get

$$\varphi(x) = \int Q(x, dy)\, \varphi(y) + h(x). \tag{3.4}$$

Write down an equation conjugate to the equation (3.4)

$$\Psi(dy) = \int \Psi(dx)\, Q(X, dy) + \nu(dy), \tag{3.5}$$

which has the explicit solution

$$\Psi(dy) = \pi(dy) \bigg/ \int s(x)\, \pi(dx) . \tag{3.6}$$

The connection between the solutions of Equations (3.4) and (3.5) is characterized by the following result. Set

$$\begin{aligned} \Psi^{(0)}(dx) &= \nu(dx), \\ \Psi^{(n+1)}(dx) &= \int \Psi^{(n)}(dy)\, Q(y, dx) + \nu(dx), \quad n = 1, 2, \ldots \end{aligned}$$

Lemma 3.1. *Let $Q(x, dy) = P(x, dy) - s(x)\nu(dy)$, where $P(x, dy)$ is the transition probability of a Markov chain which is recursive in the Harris sense satisfying (3.2) with $m_0 = 1$. Then*

a) a sequence of measures $\Psi^{(n)}(dx)$ exists with the limit at $n \to \infty$

$$\Psi^{(n)}(B) \to \Psi(B)$$

for any $B \in \mathcal{A}$, with $\Psi(dy)$ defined by formula (3.6)

b) if $h(x)$ is a measurable on $(\mathcal{X}, \mathcal{A})$ function satisfying (3.1), then

$$\lim_{n\to\infty} \int \varphi^{(n)}(x)\,\nu(dx) = \int \pi(dx)\,h(x) \Big/ \int \pi(dx)\,s(x)\,.$$

Proof. a) Set $\tilde{\Psi}^{(0)}(dy) = \nu(dy)$, $\tilde{\Psi}^{(n+1)}(dy) = \int \tilde{\Psi}^{(n)}(dx)\,P(x, dy)$, $n \ldots$ By definition, $\tilde{\Psi}^{(n)}(dy)$ is the distribution of a random variable X_n where $X_0, X_1, \ldots$ is a Markov chain with initial distribution $\tilde{\Psi}^{(0)}(dy)$ and transition probability $P(x, dy)$. By virtue of chain ergodicity $\{X_n;\ n \geq 0\}$, we have $\tilde{\Psi}^{(n)}(B) \to \pi(B)$ for any $B \in \mathcal{A}$.

Set

$$\begin{aligned}
&\varepsilon = \inf(\tilde{\tilde{\Psi}}, s),\\
&\tilde{\tilde{\Psi}}^{(n)}(dy) = \tilde{\Psi}^{(n)}(dy) \Big/ \int s(x)\,\pi(dx)\,, \qquad n = 0, 1, \ldots
\end{aligned}$$

Note that $\Psi^{(n)} \leq \tilde{\tilde{\Psi}}^{(n)}/\varepsilon$, as

$$\begin{aligned}
&\tilde{\tilde{\Psi}}^{(0)}(dy) = \nu(dy) \Big/ \int \pi(dx)\,s(x)\,,\\
&\tilde{\tilde{\Psi}}^{(n+1)}(dy) = \int \tilde{\tilde{\Psi}}^{(n)}(dx)\,\{Q(x, dy) + s(x)\nu(dy)\} \geq\\
&\geq \int \tilde{\tilde{\Psi}}^{(n)}(dx)\,Q(x, dy) + \varepsilon\nu(dy), \qquad n = 0, 1, \ldots
\end{aligned}$$

Consequently, since $\varepsilon > 0$ and the equations $xP = x$ for $(x, s) = 1$ and $x = Qx - \nu$ are equivalent, then the last equation has a unique solution and

$$\Psi^{(n)}(dy) \to \pi(dy) \Big/ \int s(x)\,\pi(dx)\,.$$

b) Let $Q^{(k)}(x, dy)$ be the Q k-th iteration

$$Q^{(k+1)}(x, dy) = \int Q^{(k)}(x, dz)\,Q(z, dy), \qquad k = 1, 2, \ldots$$

Then

$$\varphi^{(n)}(x) = \sum_{k=1}^{n-1} \int h(y)\, Q^{(k)}(x, dy) + h(x),$$

$$\Psi^{(n)}(dy) = \sum_{k=1}^{n-1} \int \nu(dx)\, Q^{(k)}(x, dy) + \nu(dy),$$

Multiplying both sides of the first equality by $\nu(dx)$, and of the second by $h(y)$, by virtue of Fubbini's theorem, we get

$$\sum \varphi^{(n)}(y)\, \nu(dy) = \sum \Psi^{(n)}(dy)\, h(y).$$

Passing to the limit at $n \to \infty$, we have

$$\lim_{n\to\infty} \int \varphi^{(n)}(y)\, \nu(dy) = \int \Psi(dy)\, h(y).$$

Q.E.D.

From Lemma 3.1, it follows that

$$EY(1) = \int \varphi(x)\, \nu(dx) = \int \Psi(dx)\, h(x) = \int \pi(dx)\, h(x) \Big/ \int \pi(dx)\, s(x).$$

By substitution $h(x) \equiv 1$, we find

$$Eb(1) = \int \pi(dx) \Big/ \int \pi(dx)\, s(x) = 1 \Big/ \int \pi(dx)\, s(x)\,.$$

With probability equal to 1,

$$\tilde{J}_{3(k)} = \left(\sum_{i=1}^{k} Y(i)/k\right) \Big/ \left(\sum_{i=1}^{k} b(i)/k\right) \xrightarrow[n\to\infty]{} \frac{EY(1)}{Eb(1)} = \int \pi(dx)\, h(x) = J_{st}.$$

Hence, $\tilde{J}_{3(k)}$ is a strongly consistent estimator of J_{st}.

Squaring both sides of (3.3) and passing to the limit, we have

$$E\bar{\varphi}^2(x) = \int E\bar{\varphi}^2(y)\, Q(x, dy) + 2h(x)\varphi(x) - h^2(x),$$

$$E\bar{\varphi}^2(x) = \int E\bar{\varphi}^2(y)\, \nu(dx) = \int \Psi(dx)\, \{2h(x)\varphi(x) - h^2(x)\} =$$
$$= (\pi,\ 2h\varphi - h^2) \,/\, (\pi, s).$$

Set $\tilde{Y}(1) = Y(1) - J_{st}b(1)$. Then

$$E\tilde{Y}(1) = 0,$$
$$D\tilde{Y}(1) = DY(1) - 2\,\mathrm{Cov}\,(Y(1), b(1))\, J_{st} + Db(1)J_{st}^2 =$$
$$= EY^2(1) - 2J_{st}E\,(Y(1), b(1)) + J_{st}^2 Eb^2(1).$$

Assuming $h(x) = \bar{h}(x) \equiv 1$ from the preceding, we find that

$$E\alpha_1^2 = (\pi,\ 2\varphi_l - \bar{h})\,/\,(\pi, s),$$

where φ_l is the solution of the equation $\varphi_l = Q\varphi_l + h$.

Let

$$\bar{\varphi}_l(x) = \sum_{i=0}^{\alpha_1} \bar{h}(X_i), \quad \bar{\varphi}_l^{(n)}(x) = \sum_{i=0}^{n_1} \bar{h}(X_i)$$

under the condition $X_0 = x$. Then

$$\bar{\varphi}_l^{(n+1)}(x)\,\bar{\varphi}^{(n+1)}(x) = \left(\int \zeta(x, dy)\bar{\varphi}_l^{(n)}(y) + h(x)\right)\left(\int \zeta(x, dy)\bar{\varphi}^{(n)}(y) + h(x)\right),$$

where $\zeta(x, dy)$ is the same as in (3.3). Passing to the limit at $n \to \infty$ and using the conjugate equation, we come to the equality

$$EY(1)b(1) = \int \nu(dx)\,\varphi_l(x)\,\varphi(x) =$$
$$= \int \pi(dx)\,\{h(x)\,\varphi_l(x) + \varphi(x) - h(x)\} \Big/ \int s(x)\,\pi(dx)\,.$$

Whence

$$D\tilde{Y}(1) = (\pi,\ 2\tilde{\varphi}\tilde{h} - \tilde{h}^2) \Big/ \int s(x)\,\pi(dx)\,,$$

where $\tilde{h}(x) = h(x) - J_{st}$, $\tilde{\varphi}$ is an iterative solution of the equation $\tilde{\varphi} = Q\tilde{\varphi} + \tilde{h}$.

Further similarly to the proof of Theorem 4.3 we obtain that

$$\lim_{k \to \infty} kD(\tilde{J}_{3(k)} - J_{st}) = D\tilde{Y}(1) \Big/ (Eb(1))^2 = (\pi,\ 2\tilde{\varphi}\tilde{h} - \tilde{h}^2) \int s(x)\,\pi(dx)$$

and the random variable $\sqrt{k}\,(\tilde{J}_{3(k)} - J_{st})$ distribution converges to the normal. We sum up the obtained results in the form of the following theorem.

Theorem 3.2. *Let the conditions of Theorem 3.1 be fulfilled and $m_0 = 1$. Then $\tilde{J}_{3(k)}$ is a strongly consistent estimator of J_{st} and the distribution of the quantity $\sqrt{k}\,(\tilde{J}_{3(k)} - J_{st})$ converges to a normal one with parameters $(0, \sigma_1^2)$. Here $Eb(1) = 1\,/ \int s(x)\,\pi(dx)$ and the value $\sigma_1^2 Eb(1)$ defining the simulation efficiency, does not depend on the selection $\{s, v\}$ in the inequality (3.2) and is equal to*

$$\big(DY(1) - 2J_{st}\,\mathrm{Cov}\,(Y(1), b(1)) + J_{st}^2 Db(1)\big)/\,Eb(1) = (\pi,\ 2\tilde{\varphi}\tilde{h} - \tilde{h}^2).$$

4. Branching paths technique

Let $\{X_n;\ n \geq 0\}$ be an ergodic Markov chain with X_n having values in $(\mathcal{X}, \mathcal{A})$, a transition probability function $P(x, dy)$, and a stationary distribution $\pi(dx)$ meeting the condition (3.2) with $m_0 = 1$.

Let $h(x)$ satisfy the conditions of Theorem 3.1. Set $X_0 \sim \nu(dx)$. Let η_n copies of the chain $\{X_n,\ Z_n\}\ :\ \{X_n^{(j)},\ Z_n^{(j)};\ n \geq 0,\ j \in 1 : \eta_n\}$ be simulated at the n–th step.

Let a measurable function $\beta(x) > 0$ be given at $(\mathcal{X}, \mathcal{A})$, $\sup \beta(x) < \infty$. Determine η_0 as a random variable

$$\eta_0 = \begin{cases} \lfloor \beta(x) \rfloor & \text{with probability } \overline{1 - \beta(x)}, \\ \lfloor \beta(x) \rfloor + 1 & \text{with probability } \overline{\beta(x)}, \end{cases}$$

where $\bar{a}$ is a whole part and $\lfloor a \rfloor$ is a fractional part of a. In the transition of the chain $X_n^{(j)}$ $(j = 1, 2, \ldots, \eta_n)$ from x to y, if $r(x, y) > 1$ add $r(x, y) - 1$ chains beginning with $X_n = X_n^{(j)}$ to a number of simulated chains, where

$$r(x, y) = \begin{cases} \lfloor \beta(y)/\beta(x) \rfloor & \text{with probability } 1 - \overline{\beta(y)/\beta(x)}, \\ \lfloor \beta(y)/\beta(x) \rfloor + 1 & \text{with probability } \overline{\beta(y)/\beta(x)}. \end{cases}$$

After simulation of η_0 chains, all those formed and those remaining should be enumerated again.

If $r(x, y) \leq 1$, then the chain simulation is discontinued with probability $1 - \beta(y)/\beta(x)$ and continued with probability $\beta(y)/\beta(x)$. If simulation is not stopped as a result of the described procedure, then chains are simulated until they jump into set $\mathcal{X} \otimes \{1\}$.

Call the quantities $\{X_n^{(\gamma)};\ \gamma \in 1 : \eta_n\}$ collected as a result of the described process which is broken with probability 1, since, as it will be shown,

$$E \sum_{n=0}^{\infty} \eta_n < \infty,$$

a cycle.

Consider the results of the simulation of k cycles. Denote them by

$$\left\{ X_n^{(\gamma)}(1);\ \gamma \in \eta_n(1) \right\}, \ldots, \left\{ X_n^{(\gamma)}(k);\ \gamma \in \eta_n(k) \right\}.$$

Make up an estimator $\tilde{J}_{3\beta(k)}$ of the quantity J_{st}

$$\tilde{J}_{3\beta(k)} = \frac{1}{k} \sum_{i=1}^{k} Y_\beta(i) \ / \ \frac{1}{k} b_\beta(i),$$

where

$$Y_\beta(i) = \sum_{n=0}^{\infty} \sum_{\gamma=1}^{\eta_n} h\left(X_n^{(\gamma)}(i)\right) / \beta\left(X_n^{(\gamma)}(i)\right),$$

$$b_\beta(i) = \sum_{n=0}^{\infty} \sum_{\gamma=1}^{\eta_n} 1 / \beta\left(X_n^{(\gamma)}(i)\right).$$

Since $Y_\beta(i)$ and $b_\beta(i)$ are IEDR, according to the Strong Large Numbers Law we obtain that for $EY_\beta(1),\ Eb_\beta(1) < \infty$

$$\tilde{J}_{3\beta(k)} \to EY_\beta(1) / Eb_\beta(1)$$

with probability equal to 1.

Subsequent calculations are analogous with the calculations in Section 3. Set

$$\bar{\varphi}_\beta(x) = \sum_{i=0}^{\alpha} \sum_{\gamma=1}^{\eta_i} h\left(X_i^{(\gamma)}\right) / \beta\left(X_i^{(\gamma)}\right), \quad \varphi_\beta(x) = E\bar{\varphi}_\beta(x),$$

$$\bar{\varphi}_\beta^{(n)}(x) = \sum_{i=0}^{n_1} \sum_{\gamma=1}^{\eta_i} h\left(X_i^{(\gamma)}\right) / \beta\left(X_i^{(\gamma)}\right), \quad \varphi_\beta^{(n)}(x) = E\bar{\varphi}_\beta^{(n)}(x),$$

where $\alpha = \min\{k \geq 0,\ \eta_k = 0\}$, $n_1 = \min\{n, \alpha\}$, $X_0^{(1)} = X$, $\eta_0 = 1$.

Evidently, $\varphi_\beta^{(n)}(x)$ does not decrease with respect to n and

$$\bar{\varphi}_\beta^{(n+1)}(x) = \int \zeta(x, dy) \sum_{\gamma=1}^{r(x,y)} \bar{\varphi}_\beta^{(n,\gamma)}(y) + h(x)/\beta(x), \tag{4.1}$$

where $\bar{\varphi}_\beta^{(n,\gamma)}(y)$, $\gamma = 1, 2, \ldots$, $r(x,y)$ are IEDR distributed by the same law as $\bar{\varphi}_\beta^{(n)}(y)$ ($\zeta(x, dy)$ is the same as in (3.3)).

According to Wald's lemma for IEDR $V_1, V_2, \ldots$ and the integer random variable η such that $P\{\eta > k \mid V_{k+1}, V_{k+2}, \ldots\} = P\{\eta > k\}$, we have

$$E \sum_{k=1}^{\eta} V_k = E\eta\, EV_1,$$

$$\mathcal{D} \sum_{k=1}^{\eta} V_k = E\eta\, \mathcal{D}V_1 + \mathcal{D}\eta (EV_1)^2.$$

Evaluating the expectations of both sides of equation (4.1) and passing to the limit at $n \to \infty$, we have

$$\begin{aligned} \varphi_\beta(x) &= \int \frac{\beta(y)}{\beta(x)} Q(x, dy)\, \varphi_\beta(y) + h(x)/\beta(x), \\ \beta(x)\varphi_\beta(x) &= \int Q(x, dy)\, \beta(y)\, \varphi_\beta(y) + h(x). \end{aligned} \tag{4.2}$$

Using conjugate equation (3.5), we find

$$E\varphi_\beta(1) = \int \nu(dx)\,\beta(x)\,\varphi_\beta(x) = \int \pi(dx)h(x) \Big/ \int \pi(dx)s(x) = EY(1).$$

Note that

$$\mathcal{D}\eta_0 = \overline{\beta(x)} - \overline{\beta(x)}^2, \quad \mathcal{D}r(x,y) = \overline{\beta(y)/\beta(x)} - \overline{\beta(y)/\beta(x)}^2.$$

Hence, according to Wald's lemma,

$$\mathcal{D}\sum_{\gamma=1}^{r(x,y)} \bar{\varphi}_\beta^{(n,\gamma)}(y) =$$
$$= \left(\beta(y)/\beta(x)\,\mathcal{D}\varphi_\beta^{(n)}(y) + \overline{\beta(y)/\beta(x)} - \overline{\beta(y)/\beta(x)}^2\right)\varphi_\beta^2(y).$$

Designate $\tilde{\varphi}_\beta(x) = \beta(x)\,\bar{\varphi}_\beta(x)$. Then, by virtue of (4.2), $E\tilde{\varphi}_\beta(x) = \varphi(x)$.

Calculating the expectations of the squares of both sides of (4.1) and passing to the limit at $n \to \infty$, we have

$$E\tilde{\varphi}_\beta^2(x) = \int \frac{\beta(x)}{\beta(y)}\,Q(x,dy)\,E\tilde{\varphi}_\beta^2(y) +$$
$$+ \int Q(x,dy)\,\varphi^2(y)\left\{1 - \frac{\beta(x)}{\beta(y)} + \frac{\beta^2(x)}{\beta^2(y)}\left(\overline{\beta(y)/\beta(x)} - \overline{\beta(y)/\beta(x)}^2\right)\right\} +$$
$$+ 2h(x)\,\varphi(x) - h^2(x).$$

Denote

$$u(x,y) = 1 - \beta(x)/\beta(y) + \frac{\beta^2(x)}{\beta^2(y)}\left(\overline{\beta(y)/\beta(x)} - \overline{\beta(y)/\beta(x)}^2\right),$$
$$u(x) = 1 - \beta^{-1}(x) + \beta^{-2}(x)\left(\overline{\beta(x)} - \overline{\beta(x)}^2\right).$$

Since

$$\{Y^2(1) \mid x\} = \left(\sum_{\gamma=1}^{\eta_0} \bar{\varphi}_\beta^{(\gamma)}(x)\right)^2,$$

where $\bar{\varphi}_\beta^{(\gamma)}(x),\ \gamma = 1,\dots,\eta_0$ is IEDR with the same distribution as $\bar{\varphi}_\beta(x)$, then

$$EY^2(1) = \int \nu(dx)\,\left\{\beta(x)E\bar{\varphi}_\beta^2(x) + \beta^2(x)\varphi_\beta^2(x) - \beta(x)\varphi_\beta^2(x) +\right.$$
$$\left. + \left(\overline{\beta(x)} - \overline{\beta(x)}^2\right)\varphi_\beta^2(x)\right\} = \int \nu(dx)\,\beta^{-1}(x)\,E\tilde{\varphi}_\beta^2(x) + \int \nu(dx)\,u(x)\,\varphi^2(x).$$

The first addend is calculated by means of the conjugate equation (3.5) and is equal to

$$\int \nu(dx)\,\beta^{-1}(x)\,E\tilde{\varphi}_\beta^2(x) = [\int \pi(dx)\,s(x)]^{-1}\times$$
$$\times \int \pi(dx)\,\beta^{-1}\left\{2h(x)\varphi(x) - h^2(x) + \int Q(x,dy)\,\varphi^2(y)\,u(x,y)\right\}.$$

Designate $\tilde{\varphi} = \varphi - J_{st}\varphi_1$, where φ_1 is the solution of the equation $\varphi_1 = Q\varphi_1 + \bar{h}$, $\bar{h}(x) \equiv 1$. Then $\tilde{\varphi}$ is the solution of $\tilde{\varphi} = Q\tilde{\varphi} + h$.

As in Section 3, we have

$$\sigma_{1\beta}^2 \stackrel{\text{def}}{=} k\mathcal{D}\tilde{J}_{3\beta(k)} \to$$
$$\to \left\{EY_\beta^2(1) - 2J_{st}E\left(Y_\beta(1)\,b_\beta(1)\right) + Eb_\beta^2(1)\right\} \Big/ \left(Eb_\beta(1)\right)^2.$$

As in the preceding (substituting $h(x) \equiv 1$ to $h(x)$), we find that

$$Eb_\beta(1) = Eb(1) = \left[\int \pi(dx)\,s(x)\right]^{-1},$$
$$\sigma_{1\beta}^2 = \left\{\int \pi(dx)\,\beta^{-1}(x)\,\left[2\tilde{h}(x)\,\tilde{\varphi}(x) - \tilde{h}^2(x) + \right.\right.$$
$$\left.\left. + \int Q(x,dy)\,\tilde{\varphi}^2(y)\,u(x,y)\right] + \int \nu(dx)\,\varphi^2(x)\,u(x)\right\} \int \pi(dx)\,s(x).$$

Let $\alpha_\beta(j) = \sum_{i=1}^{j} b_\beta(i)$. Together with the estimator $\tilde{J}_{3\beta(k)}$, consider an analogue of the estimator $\tilde{J}_{1(N)}$:

$$\tilde{J}_{1\beta(N)} = \frac{1}{N}\sum_{j=1}^{k(N)}\sum_{i=0}^{b_\beta}\sum_{j=0}^{\eta_i(j)} h\left(X_i^{(j)}(j)\right)\Big/\beta\left(X_i^{(j)}(j)\right) +$$
$$+\frac{1}{N}\sum_{i=0}^{N_1}\sum_{\gamma=0}^{\eta_i(l)} h\left(X_i^{(j)}(l)\right)\Big/\beta\left(X_i^{(j)}(l)\right),$$

where

$$k(N) = \min\{i \ge 0;\ \alpha_\beta(i) > N\} - 1,\quad l = k(N)+1,\quad N_1 = n - \alpha_\beta(l-1).$$

Note that

$$k\tilde{J}_{3\beta(k)} \le N\tilde{J}_{1\beta(N)} \le (k+1)\tilde{J}_{3\beta(k+1)},$$

where

$$k = k(N); \quad k(N)/N \to 1/Eb_\beta(1) = \int s(x)\,\pi(dx) \quad (N \to \infty).$$

Hence, $\tilde{J}_{1\beta(N)} \to J_{st}$ with probability 1 and

$$\lim_{N\to\infty} N\mathcal{D}\tilde{J}_{1\beta(N)} = \int s(x)\,\pi(dx) \lim_{k\to\infty} k\mathcal{D}\tilde{J}_{3\beta(k)}$$

does not depend on the choice of $\{s,\nu\}$.

Find the sum expectation of the number of steps for all simulated chains at one cycle (denote it by $ET(\beta)$). Substituting $h(x) = \beta(x)$, we have

$$ET(\beta) = E\sum_{n=0}^{\alpha} \eta_n = \sum_{n=0}^{\alpha}\sum_{\gamma=1}^{\eta_n} \beta\left(X_n^{(\gamma)}\right)\beta\left(X_n^{(\gamma)}\right) = \frac{\int \beta(x)\,\pi(dx)}{\int s(x)\,\pi(dx)}.$$

Therefore

$$\sigma_{1\beta}^2 ET(\beta) = L\int \pi(dx)\,\beta(x),$$

where

$$L = \int \pi(dx)\,\beta^{-1}(x)\left\{2\tilde{\varphi}(x)\,\tilde{h}^2(x) + \int Q(x,dy)\,\tilde{\varphi}^2(y)\,u(x,y)\right\} + \\ + \int \nu(dx)\,\tilde{\varphi}^2(x)\,u(x) \tag{4.3}$$

and does not depend on the choice of $\{s,\nu\}$ satisfying inequality (3.2).

Thus, we have obtained the following result.

Theorem 4.1. *Let the conditions of Theorem 3.1 be valid and $m_0 = 1$. Then the estimator $\tilde{J}_{3\beta(k)}$ defined for any measurable function $\beta(x) \geq 0$ in (4.1) is strongly consistent and the distribution of the quantity $\sqrt{k}\,(\tilde{J}_{3\beta(k)} - J_{st})$ converging to normal distribution with the variance $\sigma_{1\beta}^2$. Besides the quantity $\sigma_{1\beta}^2 ET(\beta)$ characterizing the simulation efficiency, is equal to*

$$ET(\beta)\sigma_{1\beta}^2 = L\int \pi(dx)\,\beta(x) = \left[E\sum_{n=0}^{\alpha}\eta_n \,/\, (Eb_\beta(1))^2\right] \times \\ \times \left[EY_\beta^2(1) - 2J_{st}E\left(Y_\beta(1)b_\beta(1)\right) + J_{st}^2 Eb_\beta^2(1)\right],$$

where L is defined in (4.3) and does not depend on the choice of $\{s,\nu\}$ in relation (3.2).

In the following section, this result is specified for the case when $\{X_n;\ n \geq 0\}$ is a waiting process in the queueing system of the class $GI/G/1$. In this connection, an efficient way of choosing a function $\beta(x)$ is shown.

5. Comparison of branching and importance sampling techniques

It is well known that the evaluation of the probability of rare events for queueing systems on the basis of direct simulation requires much computer time [66]. One of the most important problems of such a kind consisting the following. Consider a queueing system of the type $GI/G/1/\infty$. The problem is to estimate the probability for wainting time to be more than a given value. This problem has valuable interpretations in a series of related fields (the power evaluation of the sequential test of hypotheses, the probability evaluation of ruining [6,40, vol.2, ch.XII,97]) and may be considered as a model problem too. One elaborated technique of solving this problem on the basis of simulation is the importance sampling technique (IST) [97]. But this method requires a changed (with respect to the initial one) simulating a random process, which is not always feasible, and additional work in algorithmization and programming. A promising alternative to the IST is the branching paths technique (BPT). This technique can be easily realized and is rather universal, since it uses only the original process paths.

Simulation efficiency can be evaluated by the quantity $R = \theta^2/(ET\mathcal{D}\theta)$, where θ is the evaluated parameter value, $\mathcal{D}$ is the one-path estimator variance, and ET is the mean length of the simulated path. The quantity R describes the relative accuracy of the result for fixed expenditures as well as the value of the expenditures necessary to obtain a result with a given relative accuracy. When evaluating the probability of rare events the behavior of $R(\theta)$ for $\theta \to 0$ is of interest. The direct simulation technique is characterized by the relation $R(\theta) = O(\theta \ln^{-1}(1/\theta))$, and the IST by $R = O(\ln^{-1}(1/\theta))$. For the BPT, $R(\theta) = O(\ln^{-3}(1/\theta))$. A rigorous formulation and proof of these facts are given in this section. The presentation is based on the results given in [84,85].

5.1. Statement of the problem.

Let there be a queueing system of the $GI/G/1/\infty$ type.It is a system with one server. At its entry at the instances $t_1, t_2, \dots, t_n, \dots$ requests would arrive. Requests are maintained in order of arrival during the time $u_1, u_2, \dots, u_n, \dots$, respectively. If the device is busy, a queue of an unbounded length can be then expected. It is assumed that $X_1, X_2, \dots, X_n, \dots$ (where $X_i = U_i - V_i$, $V_i = t_{i+1} - t_i$, $i = 1, 2, \dots$) are independent, equally distributed random variables with probability density f, $EX_1 < 0$. Denote by $W_1, W_2, \dots, W_n, \dots$ the time spent by the first, second ..., n-th request in the queue. Let us consider that, at the initial instance $t = t_0 = 0$, the system is free. Then

$$W_1 = 0, \quad W_{n+1} = \max(0, W_n + X_n), \quad n = 1, 2, \dots$$

Let $S_1 = 0$, $S_{n+1} = S_n + X_n$, $n = 1, 2, \dots$; $M_n = \max\{S_1, \dots, S_n\}$. The process $\{S_n\}$ is the process of random walk related to the waiting process $\{W_n\}$. In monograph [40, vol. 2, ch. XII] one can find the main probabilistic statements which will be formulated and used further in our discussion. It is known that W_n and M_n have the same distribution and $W_n \to W$, $M_n \to M$ with respect to distribution (with the assumption that $EX_1 < 0$). Designate $\theta = P\{W \geq x\} = P\{M \geq x\} = M(x)$. An explicit expression for θ is unknown in the general case, and the problem is to evaluate θ according to the results of the simulation of $\{W_n\}$ or $\{M_n\}$.

5.2. The parameter estimator.
Define two estimates of θ: $\hat{\theta}^{(1)} = \sum_{i=1}^{M} \hat{\theta}_i^{(1)}/m$, $\hat{\theta}_i^{(1)} = 1$ if $S_n^{(i)} \in [x, \infty)$ for some n, and $\hat{\theta}_i^{(1)} = 0$ otherwise, where $S_1^{(i)}, S_2^{(i)}, \ldots, S_n^{(i)}, \ldots$ are independent paths of the process $\{S_n\}$, m is the simulated paths number

$$\hat{\theta}^{(2)} = \left(\frac{1}{m}\sum_{i=1}^{m} b_i^{(1)}\right) \Big/ \left(\frac{1}{m}\sum_{i=1}^{m} b_i^{(2)}\right),$$

$$b_i^{(1)} = \sum_{n=n_{i-1}}^{n_i-1} \chi(W_n \geq x), \quad b_i^{(2)} = n_i - n_{i-1},$$

where $n_0 = 0$, n_i is the instance of the i-th return to the initial state of the path $\{W_n\}$, and χ is the corresponding event indicator.

Note that the return to the zero state time of the path $\{W_n\}$ has a finite expectation, therefore the estimator $\hat{\theta}^{(2)}$ has been correctly defined.

The $\hat{\theta}^{(1)}$ evaluation requires a path of an infinite length to be simulated and is not applied directly. In the following section (see also [97]) it will be shown that when transferring to the changed process $\tilde{S}_i$, on the basis of $\hat{\theta}^{(1)}$ an estimator can be constructed, which requires only the paths of the finite length.

The estimator $\hat{\theta}^{(2)}$ is the ratio estimator of the regeneration technique, its properties having been investigated in [22]. On the basis of $\hat{\theta}^{(2)}$, we can construct a class of BPT estimators possessing the properties of asymptotic nonbias and consistency, and having a high efficiency. By $\hat{\theta}^{(2)}$, we can also construct estimates of IST.

5.3. Estimator $\hat{\theta}^{(1)}$ modification: the importance sampling technique.
Define the process $\{\tilde{S}_n\}$ as

$$\tilde{S}_1 = 0, \quad P\left\{\tilde{S}_{n+1} \in [u, u+du) \mid \tilde{S}_n = t\right\} = \tilde{f}(u,t)du = f(u-t)\,\frac{\tilde{h}(u)}{\tilde{h}(t)}du,$$

where $\tilde{h}(u) \in \mathcal{H} = \{\tilde{h}(u);\ \tilde{h}(u) > 0,\ \int_{-\infty}^{\infty} \tilde{f}(u,t)du \equiv 1\}$.

Determine the estimator

$$\hat{\theta}_h^{(1)} = \sum_{i=1}^{m} \hat{\theta}_{h,i}^{(1)} \Big/ m, \quad \hat{\theta}_{h,i}^{(1)} = h(z), \quad i = 1, 2, \ldots, m,$$

where z is defined by the relation $\{\tilde{S}_1^{(i)} < x, \ldots, \tilde{S}_{n-1}^{(i)} < x,\ \tilde{S}_n = z \geq x\}$ for some $n = n_i^*$ and $\hat{\theta}_{h,i}^{(1)} = 0$ if such n does not exist. Here $\{\hat{S}_n^{(i)}\}$ is the process $\{\tilde{S}_n\}$ path and $h(z) = \tilde{h}(0)/\tilde{h}(z)$. The next statement is valid.

Theorem 5.1. *The estimator $\hat{\theta}^{(1)}$ is unbiased for any $\tilde{h} \in \mathcal{H}$. The function*

$$\tilde{h}(z) = \tilde{h}^*(z) = \begin{cases} M(x-z), & z \le x, \\ 1, & \text{otherwise}, \end{cases}$$

belongs to the functions set $\mathcal{H}$ and $\mathcal{D}\,\hat{\theta}_h^{(1)} = 0$, $En_i^* = O(\ln(1/\theta))$ for $\tilde{h}(z) = \text{const}\ \tilde{h}^*(z)$.

This theorem and the subsequent theorems will be proved in Section 5.5.

Note that estimation $\hat{\theta}_h^{(1)}$ with $\tilde{h}(z) = \tilde{h}^*(z)$ requires a knowledge of the function $M(z)$. If $M(z)$ is known, then the simulation of the paths $\{\tilde{S}_n\}$ on a computer becomes unnecessary, since $\theta = M(x)$ is exactly determined analytically. Suppose that a positive root of the equation

$$\int_{-\infty}^{\infty} e^{\beta z} f(z)\, dz = 1$$

exists (which is equivalent to the statement $\exp(\beta z) \in \mathcal{H}$), then it is known that

$$M(z) \sim C \exp(-\beta z), \quad z \to \infty.$$

In this connection, we can expect that the estimator $\hat{\theta}_h^{(1)}$ with $\tilde{h}(z) = e^{\beta z}$ will be highly efficient. Siegmund [97] has shown that the estimator $\hat{\theta}_h^{(1)}$ with such a selection of $h(z)$ is the most efficient in the class of estimators where $h(z)$ has the form

$$\exp(-\gamma z + \Phi(\gamma)),$$

where $\Phi(\gamma)$ is a function with peculiar properties. Such a class of densities appears to be natural in problems of the power investigation of sequential hypothesis tests.

The following theorem characterizes the properties of the estimator $\hat{\theta}_h^{(1)}$ with $h(z) = \exp(\beta z)$.

Theorem 5.2. *Suppose that $\exp(\beta z) \in \mathcal{H}$. Then, for the estimator $\hat{\theta}_h^{(1)}$ with $\tilde{h}(z) = \exp(\beta z)$,*

$$\mathcal{D}\hat{\theta}_h^{(1)} = O(\theta^2), \quad En_1^* = O(\ln(1/\theta)), \quad \theta \to 0.$$

For the system $GI/M/1$

$$\mathcal{D}\hat{\theta}_h^{(1)} \sim \frac{\beta^2}{\alpha^2 - \beta^2}\,\theta^2,$$

where $\alpha^{-1} = Eu_1$.

Thus, for the estimator $\hat{\theta}_h^{(1)}$, where $\tilde{h}(z) = \exp(\beta z)$, the R value is equal to $O(1/\ln(1/\theta))$. Since β can be generally deduced from the equation $\int_{-\infty}^{\infty} \exp(\beta z) f(z)\, dz = 1$, the estimator is realizable.

Note that in the case of $\tilde{h}(z) = \exp(\beta z)$, the process $\{\tilde{S}_n\}$ is a random walk process with $\tilde{X}_1$ heading the distribution density equal to $\exp(\beta z) f(z)$. Let f_1 be the distribution density of random variable $(-v_1)$, and f_2 be the distribution density of u_1. Then

$$f(z)\exp(\beta z) = \int_{-\infty}^{\infty} f_1(z-y)\, e^{\beta(z-y)} f_2(y)\, e^{\beta y}\, dy.$$

Hence,

$$\tilde{X}_1 = \tilde{v}_1 + \tilde{u}_1,$$

where $\tilde{v}_1$ has a distribution density equal to $\tilde{f}_1(y) = f_1(y)\, e^{\beta y}/c_1$, and u_1 has a distribution density equal to $\tilde{f}_2(y) = f_2(y)\, e^{\beta y}/c_2$, where

$$c_1 = \int_{-\infty}^{0} f_1(y)\, e^{\beta y}\, dy, \quad c_2 = \int_{0}^{\infty} f_2(y)\, e^{\beta y}\, dy.$$

Here, from $\int_{-\infty}^{\infty} f(z)\exp(\beta z)\, dz = 1$, the finitness of the values c_1 and c_2 follows

$$1 = \int_{-\infty}^{\infty} f(z)\, e^{\beta z} dz = \int_{0}^{\infty} \left(\int_{-\infty}^{\infty} f_1(u)\, e^{\beta z} du \right) f_2(y)\, e^{\beta y}\, dy =$$
$$= \int_{0}^{\infty} \int_{-\infty}^{0} f_1(u)\, e^{\beta y} du\; f_2(y)\, e^{\beta y}\, dy = c_1 c_2.$$

Therefore, the realization of IST is reduced to the simulation of random variables with densities $\tilde{f}_1(y)$, $\tilde{f}_2(y)$. In the next section, we will consider a technique which does not require the simulation of random values distinct from v_1 and u_1, with the exception of those which are uniformly distributed.

5.4. Estimator $\hat{\theta}^{(2)}$ modification: path branching and importance sampling. First consider $\hat{\theta}^{(2)}$ properties. This estimator is a known ratio estimator in the regeneration technique. State the following main result. Designate

$$Y_j = b_j^{(1)}, \quad \alpha_j = b_j^{(2)}, \quad Z_j = Y_j - \theta\alpha_j, \quad \bar{Z} = \sum_{j=1}^{m} Z_j/m.$$

We have

$$\mathcal{D}Z_1 = \mathcal{D}Y_1 + \theta^2 \mathcal{D}\alpha_1 - 2\,\mathrm{Cov}\,(Y_1, \alpha_1)\theta.$$

Theorem 5.3. *The estimator $\hat{\theta}^{(2)}$ is asymptotically ($m \to \infty$) unbiased and asymptotically normally distributed with the variance $\frac{1}{m}\mathcal{D}Z_1/(E\alpha_1)^2$.*

The proof of Theorem 5.3 is reduced to the direct application of the Central Limit Theorem (see [22]).

Note that, in the expression for $\mathcal{D}Z_1$, the first addend is of the order $O(\theta\ln(1/\theta))$, $\theta \to 0$, the second of the order $O(\theta^2)$, and the third of the order $O(\theta^2)$, $E\alpha_1 = O(1)$.

Theorem 5.4. *For $Ex_1 < 0,\ \theta \to 0$, we have $\mathcal{D}\hat{\theta}^{(2)} = O(\theta \ln(1/\theta))$ and also*

$$\mathcal{D}Y_1 = O(\theta \ln(1/\theta)), \quad \mathcal{D}\alpha_1 = O(1), \quad \mathrm{Cov}\,(Y_1, \alpha_1) = O(\theta).$$

The value $\mathcal{D}Y_1$ is of the worst order. It turns out that this value can be reduced to the order $O(\theta^2 \ln^2(1/\theta))$ by means of the branching paths technique.

Let the function $\beta(z)$, such that $\beta(0) = 1,\ \beta(u) \le \beta(t)$ for $u < t,\ \beta(t) \equiv \beta(x),\ t \ge x$, be given on $[0, \infty)$. At each step, we shall simulate η_n paths $\{W_n\},\ \eta_1 = 1$. While simulating each path and when passing from $W_n = t$ to $W_{n+1} = u,\ r_{t,u} - 1$ supplementary paths starting with u are formed in the case of $u \le t$, else the path simulation is stopped with probability $1 - \beta(u)/\beta(t)$. All paths are simulated until they hit the zero state if they have not vanished in the process. The quantity $r_{t,u}$ is defined as follows. Designate $d = \lfloor \beta(u)/\beta(t) \rfloor$, where $\lfloor a \rfloor$ is the whole part of a, $q = \beta(u)/\beta(t) - d$. Define $r_{t,u}$ as random variable

$$r_{t,u} = \begin{cases} d & \text{with probability } 1 - q, \\ d+1 & \text{with probability } q. \end{cases}$$

Renumerate these paths $\alpha = 1, 2, \ldots, \eta_n$ at each step. Set

$$\tilde{\theta}_\beta^{(2)} = \left(\frac{1}{m} \sum_{i=1}^{m} \tilde{b}_{\beta,i}^{(1)} \right) \Big/ \left(\frac{1}{m} \sum_{i=1}^{m} \tilde{b}_{\beta,i}^{(2)} \right),$$

$$\tilde{b}_{\beta,i}^{(1)} = \sum_{n=1}^{\infty} \sum_{\alpha=1}^{\eta_n} \chi \left(W_n^{(i,\alpha)} \ge x \right) \ / \ \beta(x),$$

$$\tilde{b}_{\beta,i}^{(2)} = \sum_{n=1}^{\infty} 1/\beta \left(W_n^{(i,\alpha)} \right), \quad T_\beta = \sum_{n=1}^{\infty} \eta_n,$$

where $\{W_n^{(i,\alpha)}\}$ is one of the paths formed in the i-th cycle (each cycle starts with η_1, $W_1^{(i)} = 0$ and stops after all the paths break up), T_β is the number of simulated steps.

Theorem 5.5. *At any choice of the function $\beta(t)$ satisfying the described conditions, the estimator $\hat{\theta}_\beta^{(2)}$ is asymptotically ($m \to \infty$) unbiased, with $ET_\beta < \infty$. If there exist constants $\lambda_0,\ c > 0$, such that $\exp(\lambda_0 z) \in \mathcal{H}$, $c\exp(2\lambda_0 z) \in \mathcal{H}$ and $\beta(z) = \exp(\lambda_0 z)$, $z \le x$, then, for $x \to \infty$, we have $\theta \to 0$ and*

$$\mathcal{D}\tilde{\theta}_\beta^{(2)} = O\left(\theta^2 \ln^2(1/\theta)\right), \quad ET_\beta = O(\theta \ln(1/\theta)).$$

On the basis of the estimator $\hat{\theta}^{(2)}$ we can also determine the IST estimator. Let

$$\tilde{W}_1 = 0, \quad \tilde{W}_{n+1} = \max\{0,\ \tilde{W}_n + \tilde{X}_n\}$$

$$P\left\{ \tilde{W}_n + \tilde{X}_n \in [u,\ u + du) \mid \tilde{W}_n = t \right\} = \tilde{f}(u,t)du = f(u-t)\, \frac{\tilde{h}(u)}{\tilde{h}(t)}\, du,$$

$\tilde{h}(t) \in \mathcal{H}$. Define the estimator $\hat{\theta}_h^{(2)}$ as

$$\hat{\theta}_h^{(2)} = \left(\frac{1}{m}\sum_{i=1}^{m}\hat{b}_{h,i}^{(1)}\right) \Big/ \left(\frac{1}{m}\sum_{i=1}^{m}\hat{b}_{h,i}^{(2)}\right),$$

$$b_{h,i}^{(1)} = \sum_{n=n_{i-1}}^{n_i-1} \chi(\tilde{W}_n \geq x)/\tilde{h}(x), \quad \hat{b}_{h,i}^{(2)} = n_i - n_{i-1},$$

where $n_0 = 0$, n_i is the i-th return of the path $\{W_n\}$ to the zero state instance.

Theorem 5.6. *At any choice of the function $\tilde{h}(t) \in \mathcal{H}$, the estimator $\hat{\theta}_h^{(2)}$ is asymptotically ($m \to \infty$) unbiased. If $\exp(\beta z) \in \mathcal{H}$, then for $\tilde{h}(t) = \exp(\beta t)$, $t \leq x$, $\tilde{h}(t) = \exp(\beta x)$, $t \geq x$, $\theta \to 0$.*

$$\mathcal{D}\hat{\theta}_h^{(2)} = O(\theta^2 \ln^2(1/\theta)), \quad E\hat{b}_{h,i}^{(2)} = O(\theta \ln(1/\theta)).$$

5.5. Proof of the theorems.

When proving theorems, the method and results of random-walk investigation are used [40, vol.2, ch.12].

Proof of Theorem 5.1. Let $t < x$. Determine the functions

$$\Psi_x^{(n)}(t,z)dz = P\{S_2 < x, \dots, S_{n-1} < x,\ S_n \in [z,\ z+dz) \mid S_1 = t\},$$
$$\tilde{\Psi}_x^{(n)}(t,z)dz = P\{\tilde{S}_2 < x, \dots, \tilde{S}_{n-1} < x,\ \tilde{S}_n \in [z,\ z+dz) \mid \tilde{S}_1 = t\}.$$

The value $\Psi_x^{(n)}(t,z)dz$ for $z \geq x$ evaluates the probability of the fact that the process $\{S_k;\ k \geq 1\}$ jumps into $[x,\infty)$ for the first time at the n-th step and, notwithstanding, it emerges in the interval $[z,\ z+dz)$ on condition that $S_1 = t$. The value for $\tilde{\Psi}_x^{(n)}(t,z)dz$ at $z \geq x$ evaluates the same probability for the process $\{\tilde{S}_k;\ k \geq 1\}$.

Set

$$\Psi_x(t,z)dz = \sum_{n=2}^{\infty} \Psi_x^{(n)}(t,z)dz,$$

$$\tilde{\Psi}_x(t,z)dz = \sum_{n=2}^{\infty} \tilde{\Psi}_x^{(n)}(t,z)dz.$$

The series in the right sides of these equations converge, since they mean the probability of hitting $[z,\ z+dz)$ at some step.

Since

$$\Psi_x^{(n+1)}(t,z) = \int_{-\infty}^{x} \Psi_x^{(n)}(u,z)\, f(u-t)\, du,$$

$$\tilde{\Psi}_x^{(n+1)}(t,z) = \int_{-\infty}^{x} \tilde{\Psi}_x^{(n)}(u,z)\, \tilde{f}(u,t)\, du,$$

and by the definition

$$\tilde{\Psi}_x^{(2)}(t,z)dz = P\{\tilde{S}_2 \in dz \mid \tilde{S}_1 = t\} = \tilde{f}(z,t)dz = \Psi_x^{(2)}(t,z)\,\frac{\tilde{h}(z)}{\tilde{h}(t)}\,dz.$$

Then, by induction, we have

$$\tilde{\Psi}_x^{(n)}(t,z) = \Psi_x^{(n)}(t,z)\,\frac{\tilde{h}(z)}{\tilde{h}(t)}, \quad n = 2,3,\dots,$$

and, therefore,

$$\tilde{\Psi}_x(t,z) = \Psi_x(t,z)\,\frac{\tilde{h}(z)}{\tilde{h}(t)} = \Psi_x(\ ,z)\,\frac{\tilde{h}(z)}{\tilde{h}(t)}.$$

Note that

$$\theta = M(x) = P\{\sup_{0\le n\le\infty} S_n \ge x\} = \int_x^\infty \Psi_x(0,z)\,dz,$$

$$E\hat{\theta}_{h,1}^{(1)} = \int_x^\infty \tilde{\Psi}_x(0,z)\,\frac{h(z)}{h(0)}\,dz = \int_x^\infty \Psi_x(0,z)\,dz = \theta.$$

Consequently,

$$E\hat{\theta}_h^{(1)} = \sum_{i=1}^m E\hat{\theta}_{h,i}^{(1)}\,/\,m = \theta,$$

and the estimator $\hat{\theta}_h^{(1)}$ is unbiased for any $\tilde{h} \in \mathcal{H}$.

It is obvious that

$$E\left(\hat{\theta}_{h,1}^{(1)}\right)^2 \ge \left(E\,\hat{\theta}_{h,1}^{(1)}\right)^2 = \theta^2$$

where the equality takes plase if and only if $\hat{\theta}_{h,1}^{(1)} = \theta$ with probability 1. Hence, the condition $h(z) = 0$ for $z \ge x$ is sufficient for the equality $\mathcal{D}\hat{\theta}_h^{(1)} = 0$.

The following assertion takes place.

Lemma 5.1. *There exists a unique nondecreasing function $\tilde{h}(z)$ such that*

$$\tilde{h} \in \mathcal{H}, \quad \tilde{h}(0) = 1, \quad 1/\tilde{h}(z) = \theta \ \text{ for } \ z \ge x.$$

This function coincides with $M(x-z)/M(x)$ for $z < x$.

Proof. Let $\tilde{h}(z)$ be such that the conditions of Lemma 5.1 are satisfied. Then, since $\tilde{h} \in \mathcal{H}$, we can write

$$\int_{-\infty}^x \frac{\tilde{h}(z)}{\tilde{h}(t)}\,f(z-t)\,dz + \frac{C}{\tilde{h}(t)}\int_x^\infty f(z-t)\,dz = 1,$$

where $C = \tilde{h}(0)/\theta$. Multiplying both sides of this equation by $\tilde{h}(t)$ we obtain

$$\int_{-\infty}^{x} \tilde{h}(z) f(z-t)\, dz + C(1 - F(x-t)) = \tilde{h}(t).$$

Set $\Psi(z) = \tilde{h}(x-z)/C, \;\; z \geq 0$. By substitution $y = x - t, \;\; u = x - z$ we receive

$$\Psi(y) = \int_0^\infty \Psi(u) f(y-u)\, du + 1 - F(y), \quad y \geq 0.$$

The result from [70, Section 12.3] can be formulate in the following form

Lemma 5.2. *If there exists a nonnegative nonincreasing function* Ψ *such that*

$$\int_0^\infty \Psi(u) f(-u)\, du = F(0),$$

then the equation system

$$\Psi(y) = \int_0^\infty \Psi(u) f(y-u)\, du + 1 - F(y), \quad y > 0,$$
$$\rho(y) = \int_0^\infty \Psi(u) f(y-u)\, du + 1 - F(y), \quad y \geq 0$$

has the unique solution (Ψ, ρ) *such that* $\Psi(y)$ *is defined for* $y > 0$,*as nonnegative, nonincreasing,* $\rho(y)$ *is defined for* $y \leq 0$, *as nonnegative, nonincreasing, and* $\rho(y) \leq 1$. *Here* $\Psi(y) = M(y)$.

It was shown above that the function $\Psi(z)$ defined by the formula $\Psi(z) = \tilde{h}(x-z)/C$ satisfies the first equation from Lemma 5.2. Moreover, $\Psi(z)$ is defined at $z \geq 0, \;\; \Psi(z) > 0$ and $\Psi(z)$ is not increasing, since $\tilde{h}(y)$ is not decreasing. Check the relation $\int_0^\infty \Psi(u) f(-u)\, du = F(0)$. Since $\tilde{h} \in \mathcal{H}$ at any $t \in (-\infty, \infty)$, we have

$$\int_{-\infty}^\infty \frac{\tilde{h}(u)}{\tilde{h}(t)} f(u-t)\, du = 1.$$

Applying the substitution of variables $t = x, \;\; u = x - z$ and relations $\tilde{h}(0) = 1, \;\; 1/\tilde{h}(z) = \theta = M(x), \;\; z \geq x$, we have

$$\int_{-\infty}^\infty \frac{\tilde{h}(x-z)}{\tilde{h}(x)} f(-z)\, dz = 1,$$

whence

$$\int_0^\infty \frac{\tilde{h}(x-z)}{C} f(-z)\, dz + \int_{-\infty}^0 f(-z)\, dz = 1$$

and, consequently,

$$\int_0^\infty \Psi(z)\, f(-z)\, dz = F(0).$$

Finally, $\rho(y)$ is evidently nonnegative and does not increase, and

$$\rho(y) = \int_0^\infty \Psi(u)\, f(y-u)\, du + 1 - F(y) \le \int_0^\infty f(y-u)\, du + 1 - F(y) = 1.$$

By virtue of Lemma 5.2 $\Psi(y) = M(y)$ and, therefore, $\tilde{h}(t) = C\Psi(x-t) = M(x-t)/M(x)$, $t \le x$. Thus, the function $\tilde{h}(t)$ defined in Lemma 5.1 is unique and coincides with $M(x-t)/M(x)$ for $t < x$. Lemma 5.1 is complete.

By virtue of Lemma 5.1, the function $\tilde{h}(z) = \tilde{h}^*(z) = M(x-z)/M(x)$ for $z < x$, $\tilde{h}(z) = \tilde{h}^*(z) = 1/M(x) = 1/\theta$ for $z \ge x$, belongs to $\mathcal{H}$ and $\mathcal{D}\hat{\theta}_h^{(1)} = E(\hat{\theta}_h^{(1)})^2 - \theta^2 = 0$ for $\tilde{h}(z) = \tilde{h}^*(z)$.

Note that with such a selection of $\tilde{h}(z)$

$$\int_x^\infty \tilde{\Psi}_x(0,z)\, dz = M^{-1}(x) \int_x^\infty \Psi_x(0,z)\, dz = M^{-1}(x)\, M(x) = 1,$$

i.e., the probability of the process $\{\tilde{S}_n\}$ path hitting the interval $[x,\infty)$ equals 1.

Estimate the n_1^* expectation (the number of path steps before hitting $[x,\infty)$). We have

$$\begin{aligned} En_1^* &= \sum_{n=1}^\infty n \int_x^\infty \tilde{\Psi}_x^{(n)}(0,z)\, dz = \sum_{n=1}^\infty n M^{-1}(x) \int_x^\infty \Psi_x^{(n)}(0,z)\, dz \le \\ &\le \sum_{n=1}^\infty n \int_x^\infty M^{-1}(z)\, \Psi_x^{(n)}(0,z)\, dz. \end{aligned}$$

Since, for $\exp(\lambda_0 z) \in \mathcal{H}$, $\lambda_0 > 0$,

$$M(z) \sim \operatorname{const} \exp(-\lambda_0 z), \quad z \to \infty$$

(see [40, vol.2, ch. XII], then

$$\sum_{n=2}^\infty n \int_x^\infty M^{-1}(z)\, \Psi_x^{(n)}(0,z)\, dz \sim \operatorname{const} \sum_{n=2}^\infty n \int_x^\infty e^{-\lambda_0 z} \Psi_x^{(n)}(0,z)\, dz.$$

The expression in the right side of this relation is equal to En_1^* with constant accuracy for the process $\{\tilde{S}_n\}$ with $\tilde{h}(z) = \exp(\lambda_0 z)$, i.e., for the random walk process conjugate to $\{\tilde{S}_n\}$. Since $EX_1 < 0$, then [40, vol. 2, Section 12.4] this walk goes to $+\infty$, and

$En_1^* = \tilde{\Psi}(x)E\mathcal{F}_1$, where $\tilde{\Psi}(x)$ is the mean number of points in $(0, x)$, $\mathcal{F}_1$ is the the first ladder index value, and $\tilde{\Psi}(x) \sim \lambda_0 x$ $(x \to \infty)$, $E\mathcal{F}_1 < \infty$.

Hence, for the process $\{\tilde{S}_n\}$ with $\tilde{h}(z) = \tilde{h}^*(z)$, we have

$$En_1^* < \infty \quad En_1^* = O(x) = O(\ln(1/\theta)), \quad \theta \to 0.$$

Theorem 5.1 is complete.

Proof of Theorem 5.2. Let $\tilde{h}(z) = \exp(\beta z) \in \mathcal{H}$. Then $\tilde{S}_1, \ldots, \tilde{S}_n, \ldots$ is the process of a random walk conjugate to $\{S_n\}$. By virtue of the relations obtained during the proof of Theorem 5.1, we have

$$\mathcal{D}\hat{\theta}_{h,1}^{(1)} = \int_x^\infty \exp(-2\beta z)\,\tilde{\Psi}_x(0, z)\,dz - \theta^2 \leq$$
$$\leq e^{-2\beta x} \int_x^\infty \tilde{\Psi}_x(0, z)\,dz - \theta^2 \leq e^{-2\beta x}.$$

Since $\theta = M(x) \sim \text{const}\exp(-\beta x)$, $x \to \infty$, then $\mathcal{D}\hat{\theta}_{h,1}^{(1)} = O(\theta^2)$. The statement $En_1^* = O(\ln(1/\theta))$ has already been proved.

Let $H(t)$ be the distribution function of the first (the strict upper) ladder height for a random walk $S_1, \ldots, S_n, \ldots$. The probability of the fact that the k-th ladder height belongs to $[t,\, t + dt)$ and $(k+1)$-th belongs to $[z,\, z + dz)$ equals

$$\left(H^{*k}(t)\right)' H_z'(z - t)\,dt\,dz,$$

where H^{*k} is the k-fold convolution of the function H. Therefore,

$$\Psi_x(0, z)\,dz = \int_0^x \sum_{k=1}^{\infty} \left(H^{*k}(t)\right)' H_z'(z - t)\,dt\,dz = \int_0^x \Psi'(t)\,H_z'(z - t)\,dt.$$

For the system $GI/M/1/\infty$ it is known [40, vol 2, ch. XII, Section 4] that $H'(z) = (\alpha - \beta)\exp(-\alpha z)$, $\Psi'(z) = (\alpha - \beta)\,e^{-\beta z}$, $\alpha^{-1} = EU$. Consequently, since $h(z) = 1/\tilde{h}(z)$

$$\Psi_x(0, z) = (\alpha - \beta)^2 \int_0^x e^{(\alpha-\beta)t} dt\, e^{-\alpha z} = (\alpha - \beta)\,e^{-\alpha z}\left(e^{(\alpha-\beta)x} - 1\right),$$
$$E\left(\hat{\theta}_{h,1}^{(1)}\right)^2 = \int_x^\infty h(z)\,\Psi_x(0, z)\,dz = \int_x^\infty (\alpha - \beta)\,e^{-(\alpha+\beta)z}\left\{e^{(\alpha-\beta)x} - 1\right\} dz =$$
$$= \frac{\alpha - \beta}{\alpha + \beta}\,e^{-(\alpha+\beta)x}\left\{e^{(\alpha-\beta)x} - 1\right\},$$
$$\theta = E\hat{\theta}_{h,1}^{(1)} = \int_x^\infty \Psi_x(0, z)\,dz = \frac{\alpha - \beta}{\alpha}\,e^{-\alpha x}\left\{e^{(\alpha-\beta)x} - 1\right\}.$$

For $x \to \infty$, we have

$$\theta^2 \sim \left(\frac{\alpha-\beta}{\alpha}\right)^2 e^{-2\beta x},$$
$$E\left(\hat{\theta}^{(1)}_{h,1}\right)^2 \sim \frac{\alpha-\beta}{\alpha+\beta} e^{-2\beta x},$$
$$\mathcal{D}\hat{\theta}^{(1)}_{h,1} \sim \left\{\frac{\alpha-\beta}{\alpha+\beta} - \left(\frac{\alpha-\beta}{\alpha}\right)^2\right\} e^{-2\beta x} \sim \frac{\beta^2}{\alpha^2-\beta^2}\,\theta^2, \quad \theta \to 0.$$

The proof is complete.

Proof of Theorem 5.4. For $t > 0$ denote by $\tilde{\varphi}^{(n)}_x(t)$ the number of path $\{W_k,\ k \geq 1\}$ hits into the interval $[x, \infty)$ in n steps on condition that $W_1 = t,\ \ W_{k+1} = \max\{0,\ W_k + X_k\}$, $k \geq 1$ and the path breaks off when hitting zero,

$$\tilde{\varphi}^{(n)}_x(t) = \sum_{k=1}^{n_1} h_x(w_k),$$

where $n_1 = \min\{n,\ \alpha\}$, $\alpha = \min\{k \geq 1,\ W_k = 0\}$; $h_x(W_k) = 1$ if $W_k \geq x$ and 0 if $W_k < x$. Then, for $n = 1, 2, \ldots,$ the equality

$$\tilde{\varphi}^{(n+1)}_x(t) = \begin{cases} \tilde{\varphi}^{(n)}_x(W_2) + h_x(t), & \text{if } W_2 > 0, \\ h_x(t), & \text{if } W_2 = 0, \end{cases} \tag{5.1}$$

is valid.

Note that random variable $\tilde{\varphi}^{(n)}_x(t)$ has finite first and second moments if $EX_1 < \infty$.

Designate $\varphi^{(n)}_x(t) = E\tilde{\varphi}^{(n)}_x(t)$. Cumputing both sides of equality (5.1) expectation, we have

$$\begin{aligned} &\varphi^{(1)}(t) = h(t), \\ &\varphi^{(n+1)}(t) = \int_0^\infty \varphi^{(n)}(u)\, f(u-t)\, du + h(t), \quad n \geq 1, \end{aligned} \tag{5.2}$$

where $h(t) = h_x(t)$, $\varphi^{(n)}(t) = \varphi^{(n)}_x(t)$.

Consider the conjugate set of equalities

$$\Psi^{(1)}(t) = f(t),$$
$$\Psi^{(n+1)}(t) = \int_0^\infty f(t-u)\,\Psi^{(n)}(u)\, du + f(t), \quad n \geq 1.$$

The following result is a special case of Lemma 3.1.

Lemma 5.3. *Let $X_1 < 0$, $h(t) \geq 0$. Then*

1) there exists a finite limit $\Psi^{(n)}(t)$ at $n \to \infty$ and any fixed $t \geq 0$. This limit (denote it by $\Psi(t)$) equals const $\bar{M}'(t)$, $\bar{M}(t) = 1 - M(t)$;

2) if $\varphi^{(n)}$ is defined by (5.2) with an arbitrary nonnegative function $h(t)$ and $\int_0^\infty h(t)\, d\bar{M}(t) < \infty$, then

$$\lim_{n\to\infty} \int_0^\infty \varphi^{(n)}(t)\, f(t)\, dt = \int_0^\infty h(t)\, d\bar{M}(t) \text{ const}.$$

Proof. 1) Note that $\Psi^{(n)}$ can be represented as

$$\Psi^{(n)}(t)\, dt = \sum_{k=2}^{n} P\{W_2 > 0, \ldots, W_{k-1} > 0,\ \ W_k \in dt \mid W_1 = t\} =$$

$$= \sum_{k=2}^{n} P\{S_2 > 0, \ldots, S_{k-1} > 0,\ \ S_k \in dt \mid S_1 = t\}.$$

By virtue of the duality lemma from [40, vol.2, ch. XII], this value is equal to the expectation of the upper ladder heights in dt in n steps. The latter value converges to $\Psi\{dt\}$ for $n \to \infty$, where $\Psi\{dt\} = \text{const}\, d\bar{M}(t)$, const $= \Psi(\infty) < \infty$ at $EX_1 < 0$.

The existence of the $M(t)$ derivative follows from the existence of the $F(t)$ derivative.

2) Let $F^{*k}(u)$ be a k-fold convolution of the function F. Then

$$\varphi^{(n)}(u) = \sum_{k=1}^{n} \int_0^\infty h(t)\, dF^{*k}(u-t) + h(u),$$

$$\Psi^{(n)}(u) = \sum_{k=0}^{n} \int_0^\infty f(t)\, dF^{*k}(t-u) + f(u) \leq \Psi(u).$$

Hence, by virtue of the Fubbini theorem, we have (multiplying both sides of the first equality by $f(u)$, and of the second by $h(u)$ and integrating)

$$\int_0^\infty \Psi^{(n)}(u)\, h(u)\, du = \int_0^\infty \varphi^{(n)}(u)\, f(u)\, du.$$

The latter relation's left side limit is finite and equals

$$\int_0^\infty \Psi(u)\, h(u)\, du = \text{const} \int_0^\infty h(u)\, d\bar{M}(u).$$

Therefore, a limit of the right side exists and is equal to the same expression. This proves the lemma.

Note that, before hitting zero, the path $\{W_k;\ W_1 = t,\ k \geq 1\}$ coincides with the path of the random walk $\{S_k;\ S_1 = t,\ k \geq 1\}$ until it enters $(-\infty, 0)$. The latter can be partitioned into the independent segments

$$\{t, S_2, \dots, S_{k_1-1}\}, \quad \{S_{k_1}, \dots, S_{k_2-1}\}, \dots, \{S_{k_{\nu-1}}, \dots, S_{k_\nu - 1}\},$$

where $S_{k_1}, S_{k_2}, \dots, S_{k_\nu}$ are lower strict ladder heights, $S_{k_1}, \dots, S_{k_\nu - 1} > 0,\ S_{k_\nu} \leq 0$. Thus, for any n, the quantity $\varphi_x^{(n)}(t)$ is bounded from above by a value equal to the product of the mean number of random walk steps up to the drift into $(-\infty, 0]$ onto the mean number of lower ladder points at the interval $(0, t)$. The first of these quantities is finite since, by virtue of the duality lemma, the number of steps up to the drift into $(-\infty, 0]$ coincides with the number of upper ladder points, i.e., with the number of points of the improper (for $EX_1 < 0$) renewal process, and therefore it has a geometrical distribution. The second of the quantities is finite since the number of lower ladder points in $(0, t)$ is equal to the the number of renewal process points in the interval.

By virtue of the monotonicity and boundedness of $\varphi_x^{(n)}(t)$ (for $n \to \infty$ and fixed t), we have $\lim_{n\to\infty} \varphi_x^{(n)}(t) < \infty$ for any $t > 0$. Denote this limit by $\varphi_x(t)$. Passing to the (5.2) limit at $n \to \infty$, we have

$$\varphi_x(t) = \int_0^\infty \varphi_x(u)\, f(u - t)\, du + h_x(t).$$

Squaring both sides of the equality (5.1) and calculating the expectation, we find

$$E\left(\tilde{\varphi}_x^{(n+1)}(t)\right)^2 = \int_0^\infty E\left(\tilde{\varphi}_x^{(n)}(u)\right)^2 f(u - t)\, du + 2\varphi_x^{(n+1)}(t)\, h_x(t) - h_x^2(t). \tag{5.3}$$

The assertion $E\left(\tilde{\varphi}_x^{(n)}(t)\right)^2 < \text{const} < \infty$ for fixed $t < 0$ follows from the reasoning carried out while proving the boundedness of $\varphi_x^{(n)}(t)$. Passing to the limit at $n \to \infty$ in the equality (5.3), we have

$$E\left(\tilde{\varphi}_x(t)\right)^2 = \int_0^\infty E\left(\tilde{\varphi}_x(u)\right)^2 f(u - t)\, du + 2\varphi_x(t)\, h_x(t) - h_x^2(t). \tag{5.4}$$

From the equalities (5.3) and (5.4) and Lemma 5.3, it follows that

$$\begin{aligned} E\left(\tilde{\varphi}_x(0)\right)^2 &= \int_0^\infty E\left(\tilde{\varphi}_x(u)\right)^2 f(u)\, du = \\ &= \int_0^\infty \Psi(u) \left\{2\varphi_x(u)\, h_x(u) - h_x^2(u)\right\} du. \end{aligned} \tag{5.5}$$

Now we can obtain one more integral equation for φ_x. According to [40, vol.2, ch. XII], let us define the function $\rho(y)$ as $\sum_{n=2}^\infty \rho_n(y)$, where

$$\rho_n(y) = P\left\{S_2 > 0, \dots, S_{n-1} > 0,\ S_n < y < 0 \mid S_1 = 0\right\}.$$

The function $\rho(y)$ is the distribution function of the first (strict) lower ladder height of the random walk $\{S_n\}$ and is the eigendistribution function for $EX_1 < 0$. Let $\Psi(z)$ be the expectation of the number of (strict) upper ladder heights smaller than z. Note that $\varphi_x(t)$ is the mean of the W_k values larger or equal to x on condition that $W_1 = t$ and the path $\{W_k;\ k \geq 1\}$ breaks off when first hitting zero. The path $\{W_k;\ k \geq 1\}$ coincides with the corresponding random walk path. By the duality lemma, the number of steps of the random walk, beginning with zero, up to the drift into $(-\infty, 0]$ for some $S_k \geq y$, equals the number of upper (strict) ladder points with coordinates no smaller than y. Write down this number as $\Psi(\infty) - \Psi(y)$. It is known [40, vol. 2, ch. XII, Section 3] that $\Psi(\infty) - \Psi(y) = \Psi(\infty)\, M(y)$. Evidently, $\Psi(\infty) = \varphi_0(0)$. Let $x > t$. Placing the origin of the coordinates at the point t, we obtain that the number of values $W_k > x - t$ for $W_1 = 0$ up to the first hitting $(-\infty, 0]$, equals $\Psi(\infty) - \Psi(x-t) = \varphi_0(0)\, M(x-t)$. For $x \leq t$, this number equals $\varphi_0(0)$. Further, the difference between $\varphi_x(t)$ and the number above equals

$$\int_0^t \varphi_x(y)\, P\{W_2 > t, \ldots, W_{n-1} > t,\ W_n \in t - dy \ \text{ for some } \ n\} =$$

$$= \int_0^t \varphi_x(y)\, d\rho(t-y).$$

Denote $\tilde{M}(t)$ the function $M(t)$ for the process $\{\tilde{S}_k;\ k \geq 1\}$.

Thus, the function $Z(t) = \varphi_x(t)$ meets the equation

$$Z(t) = z(t) + \int_0^t Z(y)\, d\rho\,(t-y), \tag{5.6}$$

where $z(t) = \varphi_0(0)\, \tilde{M}(x-t)$, $\tilde{M}(x-t) = C(x-t)$ for $x - t > 0$ and $\tilde{M}(x-t) = 0$, $t \geq x$ where C is a constant.

Equation (5.6) is the renewal equation which, by virtue of the boundedness of the function $z(t) = \varphi_0(0)\, \tilde{M}(x-t)$, has a unique solution bounded on the intervals

$$Z(t) = \int_0^t z(t-v)\, dH^-(v) = \varphi_0(0) \int_0^t \tilde{M}(x+v-t)\, dH^-(v),$$

where $H^-(v)$ is the renewal function corresponding to the distribution function $\rho(v)$. It is evident that $H^-(v)$ is the mean number of random walk lower strict ladder points $\{S_k,\ k \geq 1,\ S_1 = t\}$ in the interval $(0, v)$ and $\varphi_x(t) = Z(t)$. Therefore

$$\varphi_x(t) = \varphi_0(0) \int_0^t \tilde{M}(x+v-t)\, dH^-(v). \tag{5.7}$$

Let the condition $e^{\lambda_0 t} \in \mathcal{H}$ be valid, i.e., $\int_{-\infty}^{\infty} e^{\lambda_0 t} f(t)\, dt = 1$. Then

$$H^-(v) \sim \text{const}\, v, \quad \tilde{M}(v) \sim \text{const}\, e^{-\lambda_0 v}, \quad v \to \infty,$$

and, by virtue of (5.7), we have

$$\varphi_x(t) \sim \begin{cases} \text{const}\,\theta e^{\lambda_0 t}, & t \le x, \quad t, x \to \infty, \\ c_1 + (t-x)c_2, & t > x, \quad t, x \to \infty, \end{cases}$$

where c_1 and c_2 are constants.

Using formula (5.5), we have, since $\Psi(t) \sim \text{const}\exp(-\lambda_0 t)$,

$$EY^2(1) = E\left(\tilde{\varphi}_x(0)\right)^2 = \int_0^\infty \left(2\varphi_x(t) - 1\right)\Psi(t)\,dt \sim \text{const}\,\theta.$$

Now calculate the quantity

$$E\left(Y(1)\,b(1)\right) = \lim_{n\to\infty} E\left(\tilde{\varphi}_x^{(n)}(0)\,\tilde{\varphi}_x^{(n)}(0)\right).$$

Similarly, to the precedind cosideration we have

$$\varphi^{(n+1)}(t) = \int_0^\infty \varphi^{(n)}(u)\,f(u-t)\,du + h(t),$$

where

$$\varphi^{(n)}(t) = E\left(\tilde{\varphi}_x^{(n)}(t)\,\tilde{\varphi}_0(t)\right),$$
$$h(t) = h_x(t)\,\varphi_0(t) + h_0(t)\,\varphi_x - h_x(t)\,h_0(t).$$

Using Lemma 5.3, we deduce

$$E\left(Y(1)\,b(1)\right) =$$
$$= \lim_{n\to\infty} \int_0^\infty \varphi^{(n)}(u)\,f(u)\,du =$$
$$= \int_0^\infty \left\{h_x(u)\,\varphi_0(u) + h_0(u) + h_0(u)\,\varphi_x(u) - h_x(u)\,h_0(u)\right\}\Psi(u)\,du = O(\theta).$$

Calculations carried out in a similar way give

$$Eb^2(1) = O(1), \quad E\alpha(1) = O(1),$$

which completes the proof of Theorem 5.4.

Proof of Theorem 5.5. The derivation of the principal equations is similar to those in Theorem 5.4 and in Section 4.

Denote by $\tilde{\varphi}^{(n)}_{x,\beta}(t)$ the number of hittings $[x, \infty)$ by all paths generated by $\{W_k;\ k \ge 1\}$ in n steps on condition that $W_1 = t$, $\eta_1 = 1$,

$$\tilde{\varphi}^{(n)}_{x,\beta}(t) = \sum_{i=1}^{n_1} \sum_{\gamma=1}^{\eta_i} h_x\left(W_i^{(\gamma)}\right),$$

where $n_1 = \min\{n \geq 1, \alpha\}$, $\alpha = \min\{k \geq 1;\, \eta_k = 0\}$, $W_i^{(\gamma)}$ are independent realizations of the random variable W_i. It is evident that

$$\begin{aligned} &\tilde{\varphi}^{(1)}_{x,\beta}(t) = h_x(t), \\ &\tilde{\varphi}^{(n+1)}_{x,\beta}(t) = \begin{cases} \sum_{\gamma=1}^{r_t, W_2} \tilde{\varphi}^{(n,\gamma)}_{x,\beta}(W_2) + h_x(t), & W_2 > 0, \\ h_x(t), \quad W_2 = 0, \end{cases} \end{aligned} \tag{5.8}$$

where $\tilde{\varphi}^{(n,\gamma)}_{x,\beta}(W_2)$ are independent realizations of the random variable $\tilde{\varphi}^{(n)}_{x,\beta}(W_2)$. For $EX < 0$, the random variable $\tilde{\varphi}^{(n)}_{x,\beta}(t)$ has a finite expectation which will be denoted by $\varphi^{(n)}_{x,\beta}(t)$. Calculating both sides of the expectations of (5.8) by means of Wald's lemma, we have

$$\tilde{\varphi}^{(n+1)}_{x,\beta}(t) = \int_0^\infty \frac{\beta(u)}{\beta(t)}\, \tilde{\varphi}^{(n)}_{x,\beta}(u)\, f(u-t)\, du + h_x(t). \tag{5.9}$$

Since $h_x(t) = 0,\ t > x$, then $\beta(t)\, h_x(t)/\beta(x) \equiv h_x(t)$. Comparing (5.9) and (5.2), we come to the conclusion that there exists a finite limit for $\varphi^{(n)}_{x,\beta}(t)$ at $n \to \infty$ (denote it by $\varphi_{x,\beta}(t)$) and

$$\varphi_{x,\beta}(t) = \beta(x)\, \varphi_x(t)/\beta(t).$$

Consequently,

$$\begin{aligned} EY_\beta(1) &= \int_0^\infty \varphi_{x,\beta}(t)\ (\beta(t)/\beta(x))\ f(t)\, dt = \\ &= \int_0^\infty \varphi_x(t)\, f(t)\, dt = \varphi_x(0) = EY(1), \\ Eb_\beta(1) &= \int_0^\infty \varphi_0(t)\, f(t)\, dt = \varphi_0(0) = Eb(1). \end{aligned}$$

Designate

$$\tilde{\tilde{\varphi}}^{(n)}_{x,\beta}(t) = \tilde{\varphi}^{(n)}_{x,\beta}(t)\, \beta(t) \big/ \beta(x).$$

Then

$$\mathcal{D}\tilde{\tilde{\varphi}}^{(n)}_{x,\beta}(t) = \left(\mathcal{D}\tilde{\varphi}^{(n)}_{x,\beta}(t)\right) \beta^2(t) \big/ \beta^2(x).$$

Calculating the expectations of the squares of left and right sides of (5.8) and by means of the Wald's lemma, we find

$$\begin{aligned} \mathcal{D}\tilde{\tilde{\varphi}}^{(n)}_{x,\beta}(t) = \int_0^\infty \mathcal{D}\tilde{\tilde{\varphi}}^{(n)}_{x,\beta}(u)\, \frac{\beta(t)}{\beta(u^2)}\, f(u-t)\, du + \mathcal{D}\left(\tilde{\varphi}^{(n)}_x(W_2) \mid W_1 = t\right) + \\ + \int_0^\infty \left(\varphi^{(n+1)}_x(u)\right)^2 \frac{\beta^2(t)}{\beta^2(u)}\, \gamma_{u,t} f(u-t)\, du, \end{aligned} \tag{5.10}$$

where

$$\mathcal{D}\left(\tilde{\varphi}_x^{(n)}(W_2) \mid W_1 = t\right) =$$

$$= \int_0^\infty \left(\varphi_x^{(n)}(u)\right)^2 f(u-t)\,du - \left(\int_0^\infty \varphi_x^{(n)}(u)\, f(u-t)\,du\right)^2,$$

$$\gamma_{u,t} = \overline{\beta(u)/\beta(t)} - (\overline{\beta(u)/\beta(t)})^2.$$

Note that $\gamma_{u,t} \le \max_{0\le a\le 1} a(1-a) \le 1/4$,

$$\mathcal{D}\left(\tilde{\varphi}_x^{(n)}(W_2) \mid W_1 = t\right) \le \mathcal{D}(\tilde{\varphi}_x(W_2) \mid W_1 = t) =$$

$$= \int_0^\infty \varphi_x^2(u)\, f(u-t)\,du - \left(\int_0^\infty \varphi_x(u)\, f(u-t)\,du\right)^2.$$

The third addend in the right side of (5.10) does not exceed

$$(1/4)\int_0^\infty (\varphi_x(u))^2 f(u-t)\,du.$$

Thus, the function $\varphi^{(n)}(t) = \mathcal{D}\tilde{\varphi}_{x,\beta}^{(n)}(t)$ satisfies the relations

$$\varphi^{(1)}(t) = h(t),$$

$$\varphi^{(n+1)}(t) = \int_0^\infty \varphi^{(n)}(u)\,\frac{\beta(t)}{\beta(u)}\, f(u-t)\,du + h(t),$$

where

$$h(t) = \mathcal{D}\left(\tilde{\varphi}_x^{(n)}(W_2) \mid W_1 = t\right) + \int_0^\infty \left(\varphi_x^{(n+1)}(u)\right)^2 \frac{\beta^2(t)}{\beta^2(u)}\, f(u-t)\,du \le$$

$$\le \int_0^\infty \varphi_x^2(u)\, f(u-t)\,du + \frac{1}{4}\int_0^\infty (\varphi_x(u))^2\,\frac{\beta^2(t)}{\beta^2(u)}\, f(u-t)\,du.$$

By virtue of Lemma 5.3,

$$\lim_{n\to\infty}\int_0^\infty \varphi^{(n)}(t)\,\beta^{-1}(t)\, f(t)\,dt = \int_0^\infty \Psi(t)\,\beta^{-1}(t)\,h(t)\,dt,$$

where, evidently, the left side is not decreasing with respect to n. Consequently, since $\beta(0) = 1$, for any $n \ge 1$

$$\begin{aligned} \mathcal{D}\tilde{\tilde{\varphi}}_{x,\beta}^{(n+1)}(0) &= \int_0^\infty \varphi^{(n)}(u)\,\beta^{-1}(u)\, f(u)\,du + h(0) \le \\ &\le c\int_0^\infty \Psi(t)\,\beta^{-1}(t)\,h(t)\,dt + h(0). \end{aligned} \tag{5.11}$$

Let $e^{\lambda_0 t} \in \mathcal{H}$. Then, as was mentioned above, $\bar{M}'(t) \sim const\, e^{-\lambda_0 t}$. Moreover, $\Psi(t) =$ const $\bar{M}'(t)$ according to Lemma 5.3. Suppose that

$$\int_{-\infty}^{\infty} e^{2\lambda_0 t} f(t)\, dt < k < \infty,$$

$\beta(u) = \beta^*(u)$, i.e., $\beta(u) = e^{\lambda_0 u}$ for $u < x$, $\beta(u) = e^{\lambda_0 x}$ for $u \geq x$.

Substituting the estimators of $\varphi_x(u)$ into the expression for $h(t)$, we have $\theta \to 0$,

$$h(t) = e^{2\lambda_0 t}\, \mathrm{O}(\theta^2 x^2) = e^{2\lambda_0 t}\, \mathrm{O}\left(\theta^2 \ln^2(1/\theta)\right)$$

for $x \to \infty$.

Substituting this relation to inequality (5.11) gives

$$\mathcal{D}\tilde{\tilde{\varphi}}_{x,\beta}(0) = \lim_{n\to\infty} \mathcal{D}\tilde{\tilde{\varphi}}^{(n)}_{x,\beta}(0) = \mathrm{O}\left(\theta^2 \ln^2(1/\theta)\right), \quad \theta \to 0,$$

whence

$$\mathcal{D}Y_\beta(1) = \mathcal{D}\tilde{\tilde{\varphi}}_{x,\beta}(0) = \mathrm{O}\left(\theta^2 \ln^2(1/\theta)\right), \quad \theta \to 0.$$

Note that

$$\hat{\theta}_\beta = \sum_{i=1}^{m} Y_\beta(i) \Big/ \sum_{i=1}^{m} b_\beta(i),$$

where $\{Y_\beta(i),\ i = 1, 2, \ldots, m\}$, $\{b_\beta(i),\ i = 1, 2, \ldots, m\}$ are IEDR. Therefore, standard arguments for the ratio estimators of the regeneration technique (see Sections 3 and 4), show that if $EY_\beta(1)/Eb_\beta(1) = \theta$, then $\hat{\theta}_\beta$ is an asymptotically unbiased estimator and

$$\mathcal{D}\hat{\theta}_\beta = \frac{1}{m}\left[\mathcal{D}Y_\beta(1) - 2\theta\, \mathrm{Cov}\left(Y(1), b_\beta(1)\right) + \theta^2 \mathcal{D}b_\beta(1)\right] \Big/ Eb_\beta(1).$$

We have proved that, for $\theta \to 0$ $(x \to \infty)$ and $\beta(t) = \beta^*(t)$,

$$\mathcal{D}Y_\beta(1) = \mathrm{O}\left(\theta^2 \ln^2(1/\theta)\right).$$

Similar calculations show that, under the same conditions $\mathrm{Cov}\left(Y_\beta(1), b_\beta(1)\right) = \mathrm{O}(\theta)$, $\mathcal{D}b_\beta(1) = \mathrm{O}(1)$, $Eb_\beta(1) = \mathrm{O}(\ln(1/\theta))$. Besides, for any $\beta(t)$ satisfying the conditions of Theorem 5.5,

$$Eb_\beta(1) = \varphi_0(0), \quad EY_\beta(1) = \varphi_0(0)\, M(x) = \varphi_0(0)\theta.$$

This compeletes the proof of Theorem 5.5.

The proof of Theorem 5.6 can be obtained similarily to the proof of Theorem 5.4.

CHAPTER 5. GENERAL DESIGN METHODS OF SIMULATION AND REGRESSION EXPERIMENTS

1. Cubature and interpolation formulas in the system simulation theory

Let us consider the usage of such computational mathematics classical tools as interpolation and cubature formulas in connection with system simulation problems. An overview of our findings is not exhaustive but it is concerned with the main problems in this field. It is known that averaged characteristics of any model, the simulation of which is performed by a random numbers generator, are intergrals with respect to some probability measure. Algorithms of the simulation of random values, discussed in Chapter 1, realize a change of variables in this integral that leads the integration domain to a unit hypercube, and leads the measure, with respect to which integration is carried out, to the Lebesgue measure. However, in this case, the integration multiplicity (the hypercube dimension) can be as large as possible. In practice, it is necessary to confine oneself to a finite, but may be very large, multiplicity. For details, see Ermakov [29], Sobol [99].

If the corresponding characteristics are computed as an expectaion of independent model realizations, then the amount of random numbers can be different in each realization. In this case, we reckon that an integrand depends on the number of maximal variables and is independent of high multiplicity variables in some subdomains of the hypercube (where fewer random numbers were nessesary to receive the realization).

As was repeatedly noted (Chentsov [18], Sobol [99]), a rational setting of the problem concerning the calculation of infinite–to–one integrals is possible only under the supposition that the integrand dependence decreases as the number of corresponding variables increases. The theory of the calculation of infinite–to–one integrals has been less developed.

The main lines of investigation are:

1. The advancement of the theory of the Monte Carlo method for calculation of integrals with respect to a given probability measure.

2. Special quadrature formulas for calculation of integrals with respect on the Wiener, Gauss, and other measures (Yanovich [106]).

3. Determinated methods for calculation of high multiplicity integrals (Korobov [65], Sobol [99]).

Let us look at these in the sense of their application to the simulation of complicated systems.

Let the integral

$$J[f] = \int_0^1 \dots \int_0^1 f(x_1, \dots, x_s)\, dx_1 \dots dx_s, \tag{1.1}$$

be calculated, where s is not necessarily finite. One of the simplest uses of the Monte Carlo method is to evaluate $J(f)$ by means of the sum

$$J(f) \sim S_N[f] = \frac{1}{N} \sum_{k=0}^{N-1} f(\bar{\alpha}_k), \tag{1.2}$$

where $\bar{\alpha}_k = \left(\alpha_{ks+1}, \dots, \alpha_{(k+1)s}\right)$, and α_i, $i = 1, 2, \dots$ are independent realizations of random variable which is uniformly distributed on $[0,1]$. It is known that, under the dependent choice of $\bar{\alpha}_k$, errors in formula (1.2) can be essentially diminished. It is evident that for the negative correlation of the random variable $f(\bar{\alpha}_k)$, variance of the sum $S_N[f] = \frac{1}{N}\sum f(\bar{\alpha}_k)$ decreases. The simplest example (it was one of the first to appear) of this circumstancial use is the so-called antithetic variate method of Hammersly and Morton [29]. Its essence is that

$$\tilde{S}_{2N}[f] = \frac{1}{2N}\sum_{k=0}^{N-1}\Big[f\left(\alpha_{ks}, \alpha_{ks+1}, \dots, \alpha_{(k+1)s-1}\right) + \\ +f\left(1-\alpha_{ks}, 1-\alpha_{ks+1}, \dots, 1-\alpha_{(k+1)s-1}\right)\Big] \tag{1.3}$$

is used instead of $S_{2N}[f]$.

Another example is the stratified sampling method. In its simplest form, it means that a hypercube is to be divided into two subdomains of an equal volume (in the general case, into an arbitrary number of subdomains not necessarily of equal volume), and if the first point $\bar{\alpha}_0$ jumps into the first subdomain, then the second point $\bar{\alpha}_1$ is to be in the second subdomain, etc.

A detailed description and investigation of this method can be found in the work of Ermakov [29].

A more general approach is the following [29]: integral estimators essentially more general than (1.3) or (1.2) are considered. Let $Q_1, Q_2, \dots$ be a sequence of random points in a s-dimensional hypercube and $\mathcal{A}_i(Q_1, \dots, Q_m)$, $i = 1, 2, \dots, n$ be a collection of functions chosen in accordance with the requirements given below. Take the sum (cubature sum)

$$\varkappa_n[f] = \sum_{i=1}^{n} \mathcal{A}_i(Q_1, \dots, Q_m)\, f(Q_i) \tag{1.4}$$

as the integral estimator. Constructional results are obtained if $\mathcal{A}_i$ and the joint distribution of points $Q_1, \dots, Q_m$ are selected according to the following conditions:

1. For any f, from a given class F of functions, $\varkappa_n[f]$ is an unbiased estimator of the integral

$$E\,\varkappa_n[f] = \int f(x)\,dx.$$

2. For a collection of some given functions $\{\varphi_i\}_{i=1}^{r}$, the variance of $\varkappa_n[f]$ vanishes

$$D\,\varkappa_n[\varphi_i] = 0 \quad i = 1, 2, \dots, r.$$

Both the sum $\tilde{S}_{2N}[f]$ and that obtained by means of importance sampling satisfy these conditions. The first sum is exact for any odd function (if the hypercube $\{x_1, \dots, x_s : -1 \le x_i \le 1\}$ is considered), the second one is exact for some step functions.

Another convenient example for the practical use of sums of the type (1.2) is [29]

$$\tilde{S}[f] = \frac{1}{6}\left[f(X_1) + f(X_2) + f(X_3) + f(E - X_1) + f(E - X_2) + f(E - X_3)\right], \quad (1.5)$$

where

$$X_i = \left(x_1^{(i)}, \dots, x_s^{(i)}\right), \; i = 1,2,3, \; E - X_i = \left(1 - x_1^{(i)}, \dots, 1 - x_s^{(i)}\right),$$

coordinates $x_j^{(1)}$, $x_j^{(2)}$, $x_j^{(3)}$ are connected with the relation $\left(x_j^{(1)}\right)^2 + \left(x_j^{(2)}\right)^2 + \left(x_j^{(3)}\right)^2 = 1$, $j = 1, \dots, s$; they are random and uniformly distributed on the unit sphere surface. A quadrature formula corresponding to the sum $\tilde{\tilde{S}}_N[f]$ is exact for the first and second degrees of all independent variables.

Methods presented in the book by Ermakov [29] make it possible to construct and investigate the majority of sums of the type (1.2). A large class of such sums is connected with the means of random interpolation which will be discussed below.

Now suppose that a hypercube is divided by some means into N subdomains of an equal volume, the diameter of which does not exceed δ. Choosing one point X_i in each subdomain in a random way and forming the sum

$$S_N = \frac{1}{N}\sum f(X_i), \quad (1.6)$$

we have one of the stratified sampling species. If a similar choice is accomplished only once, then the difference between random and determinate samplings vanishes. Herein, we can refuse to use probability-theoretic error analysis and instead use determinate methods under the supposition that f belongs to a class narrower than the class of integrable functions. It was this approach that was developed in the works of Sobol [100] and others. Cubature formulas constructed for the calculation of high multiplicity integrals such as Korobov's formulas [65], also can be applied in system simulation, although the integrand belonging to many classes is very difficult to check. Usually, an experiment is required in order to check the expendiency of using a corresponding cubature formula.

This formula is applied in the following way. Let the cubature sum

$$S_M(g) = \frac{1}{M}\sum_{i=1}^{M} J(X_i), \quad X_i = \left(x_i^{(1)} \dots x_i^{(s)}\right) \quad (1.7)$$

be meant for an approximate calculation of integral $\int_{\mathcal{D}_s} J(x)dx$ on the unit s-dimensional hypercube $\mathcal{D}_s$. As was indicated, the problem of evaluating expectation in simulation problems is reduced to evaluating an integral of high (may be random) multiplicity r on the unit hypercube $\int_{\mathcal{D}_r} F(Y)dY$. If $s \geq r$, then the algorithm means using the numbers $x_i^{(1)}, x_i^{(2)}, \dots, x_i^{(s)}$ instead of the corresponding random numbers and performing M tests. If here some of the last numbers $x_i^{(j)}$ are "redundant", there is nothing bad in it. If they are few in number ($s \leq r$), then their shortage is completed with pseudorandom numbers. It means that a cubature formula is used with respect to the first s variables. Here may occur that

1. The result is critical concerning M. Some formulas require the number of tests to be a multiple to M.

2. Strictly speaking, empirical methods for the estimation of variance cannot be used for error estimation.

3. It may also happen that the order of error decreasing will have a higher order than when using the simple Monte Carlo method.

The described methods do not touch upon the main part of simulation algorithms. Transformations are carried out only at the level of random number generators. These numbers are replaced with dependent random and nonrandom (quasirandom) numbers.

Methods of improving efficiency, such as the importance sampling technique, analytical integration by part of the variables, and others, essentially differ from these methods. Their detailed description pertaining to the integral calculation problem can be found in [29]. As a rule, they modify the initial model and here its physical visualization is lost.

All the methods listed above are methods of experimential design for simulation. Those methods that change the initial model are techniques of special simulation experimential design. As for the random quadrature formulas, we can observe their connection using the classical methods of regression experimental design.

The classical scheme of regression that is linear with respect to the parameters, can be described as follows. Suppose that a function $f(x)$, $x \in \mathcal{X}$ can be observed with a random error in the points of $\mathcal{X} \subset R^s$, so the values

$$\xi(x_j) = f(x_j) + \varepsilon_j, \quad E\varepsilon_j = 0, \quad j = 1, \dots, \mathcal{N}, \tag{1.8}$$

are accessible for processing. Suppose also that

$$f(x) = \sum_{i=1}^{n} c_i \varphi_i(x), \tag{1.9}$$

where c_i are unknown parameters and $\varphi_i(x)$ are given linearly independent functions. Let $E\varepsilon_j^2 = \sigma^2 < \infty$ and $\mathcal{N} \geq n$. Then the covariance matrix of the estimators by means of the least-square technique for parameters c_i, is

$$\|\mathrm{Cov}\,(\hat{c}_i, \hat{c}_j)\|_{i,j=1}^{n} = \sigma^2 \left(\left\| \sum_{j=1}^{\mathcal{N}} \varphi_k(x_j)\, \varphi_l(x_j) \right\|_{k,l=1}^{n} \right)^{-1}.$$

Evidently, this matrix depends on the choice of points $x_1, \dots, x_{\mathcal{N}}$, in which the experiment is performed. A set of observation points $x_1, \dots, x_{\mathcal{N}}$ is called a regressive experiment design.

One of the most widely-used optimum criteria is the maximal value of the determinant of the matrix which is the inverse of the covariance matrix (D-optimal designs) [38].

If the simulation aim is to elucidate the dependence of model behavior on some parametres (parameter X from $\mathcal{X}$), and the supposition that the dependence has the form of (1.9) is well grounded, then we can perform simulation at the points of the, in some respect, optimal design (for example, D-optimum design) and estimate the unknown parameters c_i by the least-squares technique.

In some cases, it is advisable to use randomized experimental designs [38] when observation points are chosen at random in accordance with the density of the joint distribution

$$g(x_1, \ldots, x_N) = C \cdot \det \left\| \sum_{j=1}^{N} \varphi_k(x_j) \varphi_l(x_j) \right\|_{k,l=1}^{n}, \tag{1.10}$$

where C is the normalization constant.

If $\{\varphi_k\}$ form an orthonormal system and f is a square integrable function, not necessarily represented in the form (1.9), then for the random values $x_1, \ldots, x_N$ distributed in accordance with (1.10), the equality

$$E\,\widehat{C}_i = (\varphi_i, f) = \int f(x)\,\varphi(x)\,dx, \tag{1.11}$$

is valid. It means that the expectations of the parametres of the LST-estimators of regression are Fourier coefficients of $f(x)$ over the orthonormal function system φ_i, and $\sum_{i=1}^{n} \hat{c}_i\,\varphi_i(x)$ is an unbiased estimator of a segment of the function f Fourier series, i.e., $E \sum \hat{c}_i\,\varphi_i(x)$ approximates the function f in the sense of a mean square value in the best way. Unlike the determined setting, here we deal with the approximation of a function f by means of a generalized polynome $\sum c_i\,\varphi_i(x)$, and not with the exact equality (1.9). The case of $N = n$ is of special interest. Here the least-square technique coincides with a simple interpolation. If the function set $\{\varphi_i\}$ is such that

$$\operatorname{mes}\{x_1, \ldots, x_n : \det \|\varphi_i(x_j)\|_{i,j=1}^{n} = 0\} = 0 \tag{1.12}$$

(a regular set of functions), then the variances of all $\hat{c}_i$ are equal to each other

$$D\,\hat{c}_i = \left[\int \left(f(x) - \sum_{i=1}^{n} (f, \varphi_i)\,\varphi_i(x) \right)^2 dx \right] + \sigma^2. \tag{1.13}$$

Repeating measurements (calculations) of f at the same points, it is easy to split the variance into systematic (the first addend) and random (the second addend) components.

This technique is an important tool when checking the adequacy of the regression model.

In the general case, the inequality

$$D\,\hat{c}_i \leq \left[\int \left(f(x) - \sum(\varphi_i, f)\,\varphi_i(x)\right)^2 dx\right] + \frac{\sigma^2}{N-n+1}. \tag{1.14}$$

is valid for the variance $\hat{c}_i$. More detailed information about interpolation formulas with random nodes can be found in [29, 30]. In addition to the aforesaid aspects, it is important to point out that interpolation formulas with random nodes can be used for receiving quadrature formulas with random nodes. As was indicated, these can be used in system simulation. Equality (1.11) is valid for any integrable function f. That is, the reason for calling the designs of the experiment $x_1, \dots, x_N$ unbiased designs. Nonbiased conditions (1.11) are valid in such a wide class of functions f only for random designs. If we narrow the class of functions and require (1.11) to be valid for some finite function set $\psi_1, \psi_2, \dots, \psi_L$ $(L > m)$, then we can construct determined designs which are often called unbiased too. They are often used in system-simulation problems. The corresponding theory that generalizes the classic Kiefer and Wolfowitz results [31, 56] is given in the next section.

2. Experimental designs minimizing bias and related problems

In the classical theory of the design of regression experiments, going back to the papers of Kiefer and Wolfowitz [56, 57], the problem of selection of the conditions for the execution of experiment (x) has been considered. The results of this experiment can to be described by the standard linear regression equation

$$y_j = \Theta^T f(x_j) + \varepsilon_j,$$

where $\{y_j\}$ are observation results, $\Theta^T = (\theta_1, \dots, \theta_n)$, $f(x) = (\Psi_1(x), \dots, \Psi_n(x))^T$, $\Psi_i(x)$ are known basis functions given on $\mathcal{X}$, $\{\xi_j\}$ are errors $(j = 1, 2, \dots, N;\ i = 1, \dots, n)$, $\{\theta_i\}$ are unknown parameters ($\mathcal{X}$ is a design set with the lattice of a linear topological space). Here one of the postulates is the lack of a systematic error $(E\varepsilon_j = 0)$. Formally, the difficulties connected with this error can be avoided by introducing a rather large number of linearly independent functions forming a space basis to which the dependence under investigation a-priori belongs. Such an approach is reasonable in problems wherein the number of tests can increase without a limit. Other approaches were considered in [27, 32, 77, 95]. The results below are connected with the conception of so-called unbiased designs of the experiment introduced in [27]. In this paper, the experimental design was interpreted as a joint distribution of the observation points satisfying some special unbias conditions and the dependence under study was supposed to be an element of an infinite-dimensional space of functions, square integrable with respect to some measure (see [27, 32]). The term "unbiased design" is connected with the problems of the improvement of the efficiency of the Monte Carlo technique and, further, we will consider designs which minimize bias (MB designs), since this definition seems to be natural in the theory of experimental design.

This section deals with modification of the problem wherein the true dependence is supposed to belong to the finite-dimensional space of square integrable functions, and an approximating model is supposed to belong to this space subspace. By MB design, we will understand a collection of elements of the set $\mathcal{X}$ corresponding to the execution conditions, which minimizes the divergence of the model expectation from the true dependence (Section 2.1). Such a design proves to also provide an unbiased parameter estimator of the model, and its points coincide with the nodes of some cubature formulas. The developed apparatus enables us to reveal a similarity with the classical theory of experiments. We can prove an equivalence theorem wherein a projector matrix onto the subspace, generated by linear combinations of basis functions vector values at the design points, plays a role analogous to that of the information matrix in the equivalence theorem of Kiefer and Wolfowitz. (Section 2.2). In the case when the true dependence belongs to the space generated by algebraic polynomials the MB design can be found analytically (Section 2.3). For the general case, the designs are supposed to be found by means of a specially developed numerical method (Section 2.4).

2.1. Statement of the problem.

Let $\Psi_1(x), \Psi_2(x), \dots, \Psi_n(x)$ be the orthonormal basis of a finite-dimensional space of functions to be the square integrable with respect to the measure $\lambda(dx)$ defined on the σ–algebra of subsets of the linear topological space $\mathcal{X}$.

Denote $f^T(x) = (\Psi_1(x), \dots, \Psi_n(x))$, $\varphi^T(x) = (\Psi_1(x), \dots, \Psi_m(x))$, $m < n$. Let the experiment results (y_j) be described by a model of the linear regression

$$y_j = \Theta^T f(x_j) + \varepsilon_j, \quad j = 1, 2, \dots, N,$$

where $\Theta^T = (\theta_1, \dots, \theta_n)$ is a vector of unknown parameters, ε_j are random errors, $E\varepsilon_j = 0$, $E\varepsilon_i\varepsilon_j = \sigma^2\delta_{ij}$, δ_{ij} is the Kronecker symbol, and $x_1, \dots, x_N \in \mathcal{X}$ are observation points which are to be considered as distinct. The collection $\xi = \{x_1, \dots, x_N\}$ is called an experimental design.

Let there be a model of the type $\Theta^T_{(1)}\varphi(x)$ where $\Theta^T_{(1)} = (\theta_1, \dots, \theta_m)$ is necessary to be estimated according to the results of the experiment. Let us consider the estimators of $\Theta_{(1)}$ in the form $\widehat{\Theta}_{(1)} = AY$, where A is an arbitrary $m \times N$ matrix and $Y = (y_1, \dots, y_N)^T$ is a vector of observation results. Let $F = (\Psi_i(x_j))_{j=1,2,\dots,N,\, i=1,2,\dots,n}$ be a matrix of the order $N \times n$.

The quality of the model $\widehat{\Theta}^T_{(1)}\varphi(x)$ will be evaluated by the quantity $J = E\int_{\mathcal{X}} |\widehat{\Theta}^T_{(1)}\varphi(x) - \Theta^T f(x)|^2\, \lambda(dx) = J_1 + J_2$, where

$$J_1 = \int_{\mathcal{X}} |\varphi^T(x)AF\Theta - \Theta^T f(x)|^2 \lambda(dx),$$

$$J_2 = E\int_{\mathcal{X}} |\varphi^T(x)AF\Theta - \varphi^T(x)AY|^2 \lambda(dx).$$

The first term J_1 correponds to the systematic error and the second one J_2 to the random error.

The criterion J manifests a mean with the weight $\lambda(dx)$ divergent of the model from true dependence. In practice, a weight $\lambda(dx)$ is to be selected in accordance with an the investigator's evaluation of the significance of the value x. The investigator also selects the basis functions $\{\Psi_i(x),\ i = 1, 2, \ldots, n\}$.

Denote $T = (I_m \mid 0)$, the matrix T order is equal to $m \times N$. Therefore, $\Theta_{(1)} = T\Theta$. Let $\|v\| = \left(\sum v_i^2\right)^{1/2}$ be the Euclidean norm of the vector v, $\Theta_{(2)}^T = (\theta_{m+1}, \ldots, \theta_n)$. Since the vector components $f(x)$ are orthonormalized with respect to the measure $\lambda(dx)$, we have

$$\begin{aligned} J_1 &= \|\Theta_{(2)}\|^2 + \Theta^T (AF - T)^T (AF - T)\Theta, \\ J_2 &= \sigma^2 \ \mathrm{tr}\ AA^T. \end{aligned} \tag{2.1}$$

Let us suppose that there exists a matrix A such that $AF = T$. Illustrative examples are given below. The next lemma is valid.

Lemma 2.1. *If there exists a matrix $A = \bar{A}$ such that $\bar{A}F = T$, then the minimum of J_1 with respect to A and ξ is attained for any vector Θ if and only if $AF = T$, and equals $\|\Theta_{(2)}\|^2$.*

The lemma proof is evident from (2.1) for J_1.

Denote by A^- a matrix with the property $A^-AA^- = A^-$. By A^+ we will denote a matrix A^- which possesses complementary properties

$$AA^+A = A, \quad (AA^+) = AA^+, \quad (A^+A)^T = A^+A.$$

The matrix A^+ is uniquely defined [93] (it is called generalized inverse in the Moore–Penrose sense or, sometimes, pseudoinverse). It is known that $AF = T$ is the necessary and sufficient condition of the nonbias of the estimator AY for $\Theta_{(1)} = T\Theta$. The variance matrix of the estimator satisfying this condition has the form $\sigma^2 AA^T$ and, in the class of linear unbiased estimators [93]

$$\mathcal{D}\widehat{\Theta}_{(1)} \geq \sigma^2 \tilde{A}\tilde{A}^T, \quad \text{where} \quad \tilde{A} = T(F^TF)^-F^T,$$

is valid. Moreover, $\tilde{A}F = T$,

$$\tilde{A}\tilde{A}^T = T(F^TF)^-F^TF(F^TF)^-T^T = T(F^TF)^-T^T$$

and the matrix $T(F^TF)^-T^T$ is the same for all $(F^TF)^-$ (under a fixed F).

Therefore

$$T(F^TF)^-T^T = T(F^TF)^+T^T$$

and

$$\tilde{A}\tilde{A}^T = A^*A^{*T},$$

where $A^* = T(F^TF)^+F^T$.

Hence, the following lemma is valid

Lemma 2.2. *J_2 minimum with respect to A, satisfying the condition $AF = T$, is attained for $A = A^* = T(F^TF)^+F^T$.*

So, if there exists a matrix $\bar{A}$ such that $\bar{A}F = T$, then by virtue of Lemmas 1.1 and 1.2, $J = J_1 + J_2$ minimum with respect to A is attained for the matrix $A^* = T(F^TF)^+F^T$ (the condition $A^*F = T$ is valid for this matrix).

Definition 2.1. Let a design $\xi^* = \{x_1^*, \dots, x_N^*\}$ be called a design which minimize bias (MB design) if

$$\min_A J_1(\xi^*, A) = \min_{\xi, A} J_1(\xi, A) = \|\Theta_{(2)}\|^2.$$

Further, the properties of these designs will be studied.

From (1.1), it follows that a design is MB design if and only if $AF = T$.

2.2. Equivalence theorem.

The condition $AF = T$ may be treated as follows. Write it in the form

$$\sum_{j=1}^{N} a_{ij}\Psi_1(x_j) = \delta_{1j} = \int_{\mathcal{X}} \Psi_l(x)\,\Psi_i(x)\,\lambda(dx), \tag{2.2}$$

$i = 1, 2, \dots, m;\ \ l = 1, 2, \dots, n.$

An approximate equality of the form

$$\int_{\mathcal{X}} g(x)\,\lambda(dx) \approx \sum_{j=1}^{N} a_i\, g(x_i) \tag{2.3}$$

is called a cubature formula, where a_i are some numerical factors called the weights of the cubature formula and x_i are elements of the subspace $\mathcal{X}$ called nodes. Usually, such a choice of nodes and weights for the cubature formula is considered that makes (2.3) exact for the function g from some class (see [87]).

Thus, the condition $AF = T$ for some matrix A is equivalent to the existence of m cubature formulas which are exact for all generalized polynomials of the form $P(x) = \Theta^T g(x)$ corresponding to weight functions $\Psi_1(x)\,\lambda(dx), \dots, \Psi_n(x)\,\lambda(dx)$. This simple remark enables us to find an important relation between the evaluating a MB design problem and the theory of cubature formulas.

In some cases [95], there exist cubature formulas which are exact for products $\Psi_i(x) \times \Psi_j(x)$, $i = 1, \dots, m;\ j = 1, \dots, n$. Here, (2.1) with a matrix A of the form $F_1^T P$ is valid, where $F_1 = (\Psi_i(x_j))$, $i = 1, \dots, m;\ j = 1, \dots, N,\ \ P = \operatorname{diag}(p_1, \dots, p_N)$, $p_1, p_2, \dots, p_N$ are weights of the cubature formula and x_i, $i = 1, 2, \dots, N$ are its nodes. These nodes form the MB design.

Further, we need some auxiliary results.

Lemma 2.3. *The matrix* $S = S(\xi) = (F^TF)^+F^TF$ *corresponds to the projector from* R^n *onto the linear space spanned by the vectors* $f(x_1), \dots, f(x_N)$.

Proof. Denote $C = F^TF$. By definition of C^+, we have

$$C^+CC^+ = C^+$$

and multiplying both sides by C from the right, we have

$$(C^+C)(C^+C) = C^+C,$$

i.e., $(C^+C)^2 = C^+C$ and, thus, $S = C^+C$ is the matrix of a projector.

Let $\mathcal{L}(C)$ be the linear space spanned by the columns of the matrix C. It is known [93] that $\mathcal{L}(F^T) = \mathcal{L}(F^TF)$ and, hence, now we only have to prove that

$$(C^+C)b = b$$

if and only if b has the form

$$b = Ca \tag{2.4}$$

for a vector $a \in R^n$.

Let b have the form (2.4) then, since $C^T = C$,

$$((C^+C)b)^T = ((C^+C)Ca)^T = a^TC^T(C^+C)^T = a^TC(C^+C) = a^TC = a^TC^T = b^T,$$

hence

$$(C^+C)b = b.$$

On the contrary, let the last relation be valid. Note that $Cb \in \mathcal{L}(F^TF)$ and, therefore, $b = C^+Cb \in \mathcal{L}(F^TF)$, i.e., b has the form (2.4).

The lemma is complete.

Lemma 2.4. *For* $A = A^* = (F^TF)^+F^T$ *and any design* ξ, *the quantity* J_1 *has the form*

$$J_1 = \|\Theta_{(2)}\|^2 + \Theta^T\,(TS(\xi) - T)^T\,(TS(\xi) - T)\Theta.$$

It is sufficient to substitute $A = A^*$ into (1.1) to prove Lemma 2.4.

Lemma 2.5. *For any design* ξ, *the inequality*

$$\operatorname{tr}\, TS(\xi)T^T \le m$$

is valid and the equality is attained if and only if

$$TS(\xi)T^T = I.$$

Proof. Denote $C = F^T F$ as in the proof of Lemma 2.1. The matrix C is symmetric and, thus, can be lead to a diagonal form by some orthogonal transformation $(\mathcal{D})$

$$\mathcal{D}^T C \mathcal{D} = \Lambda, \quad C = \mathcal{D}\Lambda\mathcal{D}^T,$$

where Λ is a diagonal matrix. Consider the matrix Λ' resulting from Λ by the inversion of all the nonzero elements. Then

$$\mathcal{D}\Lambda'\mathcal{D}^T = C^+,$$

which can be checked by direct calculation. We have

$$S = C^+C = \mathcal{D}\Lambda'\mathcal{D}^T\,\mathcal{D}\Lambda\mathcal{D}^T = \mathcal{D}\Lambda'\Lambda\mathcal{D}^T.$$

Therefore, all eigenvalues of $S(\xi)$ are equal to 0 or 1 and

$$\lambda_{\max}(TST^T) \le \sup_{v^T v} v^T S(\xi) v = 1.$$

Hence, the assertion $\operatorname{tr} TS(\xi)T^T = m$ is equivalent to

$$TS(\xi)T^T = I.$$

The lemma is complete.

The next result follows from the proof of Lemma 2.5.

Lemma 2.6. *The statements*
1) tr $TS(\xi)T^T = m$,
2) $S(\xi)$ *has the form* $\begin{pmatrix} I_k & 0 \\ 0 & B \end{pmatrix}$, *where* $k \ge m, \ B \ge 0$,
3) $TS(\xi) = T$,
are equaivalent.

Since, by virtue of Lemma 2.2, $J_1 \ge \|\Theta_{(2)}\|^2$ and $J_1 = \|\Theta_{(2)}\|^2$ for any Θ if and only if $TS(\xi) = T$, then uniting the results obtained above, we will obtain a theorem analogous to the equivalence theorems of classic experimental design theory [31,58].

Theorem 2.1. *(Equivalence theorem). The following statements are equivalent:*
1) a design $\xi^* = \{x_1^*, \dots, x_N^*\}$ *is an MB design;*
2) there exists a matrix A *such that* $AF^* = T$, *where* $F^* = (\Psi_i(x_j^*))$ *(then an estimate* $\widehat{\Theta}_{(1)} = AY$ *is an unbiased estimator of* $\Theta_{(1)}$*);*
3) there exist m *cubature formulas with weight functions* $\Psi_1(x)\,\lambda(dx), \dots, \Psi_k(x)\,\lambda(dx)$ *and nodes at the design* ξ^* *points, which are exact for all generalized polynomials;*
4) tr $TS(\xi^*)T^T = m$, tr $TS(\xi)T^T \le m$ *for any design* ξ;
5) $S(\xi^*)$ *has the form* $\begin{pmatrix} I_k & 0 \\ 0 & B \end{pmatrix}$, $k \ge m$, $B \ge 0$;
6) $TS(\xi^*) = T$.

Remark 2.1. By the Sobolev theorem (87, p.78), condition 3) is equivalent to the following: ξ^* is an MB design if and only if, for any generalized polynomial $P(x) = \theta^T f(x)$ vanishing at design ξ^* points,

$$\int_{\mathcal{X}} P(x)\,\Psi_i(x)\,\lambda(dx) = 0, \quad i = 1, 2, \dots, m$$

is valid.

2.3. Minimizing bias designs for polynomial regression.

First consider the case of polynomials in one variable. Let $\mathcal{X} = [a, b]$ be a segment of R^1 and $\Psi_i(x), \ i = 0, 1, \dots, n$ be polynomials of degree $0, 1, 2, \dots, n$ orthonormalized with the weight $\lambda(dx)$ at $[a, b]$. Denote an integral part of S by $\lfloor S \rfloor$.

Theorem 2.2. *For indicated $\mathcal{X}$ and $\Psi_i(x), \ i = 0, 1, \dots, n$:*

1) With $N \geq n + 1$, every design is an MB design;

2) With $n + 1 > N \geq \lfloor \frac{n+m}{2} \rfloor + 1$, a design ξ^* is an m.b. design if and only if any one of the conditions is valid:

a) the points of the design $x_1, \dots, x_N$ are zeros of a polynomial orthogonal to all polynomials of a degree equal to $n + m - N$ or of a lesser degree;

b) there exists a cubature formula which is exact for all polynomials of a degree equal to $m + n$ or of a lesser degree with nodes at the points of this design;

3) With $N \leq \lfloor \frac{n+m}{2} \rfloor$, no MB designs exist.

Proof. 1) By definition, the points of a design are mutually dinstinct. Let us take any $n + 1$ of these points. Then $\det F$ (where $F = (\psi_i(x_j)), \ i = 0, 1, \dots, n, \ j = 1, \dots, n + 1$) is the Vandermonde determinant, hence $\det F \neq 0$. Therefore, $S(\xi) = I$ and, according to Theorem 2.1, a design ξ is an MB design.

2) By virtue of Remark 2.1, the statement that the design is an MB design is equivalent to the proposition that, for any polynomial $P(x)$ of a degree $\leq n$, $P(x) = \Theta^T f(x)$, vanishing at the design points

$$\int P(x)\,\Psi_i(x)\,\lambda(dx) = 0, \quad i = 0, 1, \dots, m \tag{2.5}$$

is valid. According to Besout's theorem, every polynomial $h_n(x)$ of a degree n vanishing at the point $x = a$ is represented by $(x - a)\,h_{n-1}(x)$, where $h_{n-1}(x)$ is a polynomial of a degree $n - 1$. Therefore, $P(x)$ has the form $P(x) = L(x)\,Q(x)$, where

$$L(x) = \prod_{i=1}^{N} (x - x_i)$$

and $Q(x)$ is a polynomial of a degree $\leq n - N$.

Thus, a polynomial $L(x)$ is orthogonal to the products of all possible polynomials of a degree $\leq n - N$ to various polynomials of a degree $\leq m$. Since orthogonal polynomials have only real roots, they are represented as products of polynomials of a lesser degree. Consequently, a set of polynomials to which the polynomial Q is orthogonal involves orthogonal polynomials of degree $0, 1, \ldots, n+m-N$. As these orthogonal polynomials form a basis in the space of polynomials of a degree $\leq n+m-n_1$, then (3.1) is equivalent to the $L(x)$ orthogonality of all polynomials of a degree $\leq n+m-N$ and, thus, 2), a) is complete.

2), b) can be similarly proved.

Hypothesis 3) is derived from the above considerations since, for $N \leq \lfloor \frac{n+m}{2} \rfloor$, a polynomial $L(x)$ cannot be orthogonal to all polynomials of a degree $\leq n+m-N$.

Theorem 2.2 is complete.

An analogue of hypothesis 2) is lacking for polynomials in many variables althought 2), b) is valid only with some complementary conditions. The representation of a polynomial by the elements of a homogeneous basis replaces Bezout's theorem. Further, we need to introduce conceptions of an ideal and of a homogeneous basis [87].

A set of polynomials (in n variables) α satisfying the conditions

1) $f \in \alpha$, $g \in \alpha$ implies $f + g \in \alpha$,

2) g is a polynomial in n variables, $f \in \alpha$ implies $fg \in \alpha$

is called a polynomial ideal.

A set of polynomials $h_1, \ldots, h_r$ such that any polynomial from α is represented as

$$a_1(x)\, h_1(x) + a_2(x)\, h_2(x) + \ldots + a_r(x)\, h_r(x),$$

where $a_1(x), a_2(x), \ldots, a_r(x)$ are polynomials in n variables, is called the basis of an ideal α. If an ideal α has a basis consisting of homogeneous polynomials, then the ideal and the pointed basis are called homogeneous.

Let $\Psi_0(x), \ldots, \Psi_n(x)$ be orthonormalized polynomials in k variables of a degree $\leq d_2$, $n+1$ be equal to the number of linearly independent monomials of a degree $\leq d_2$: $n+1 = M(d_2, k) = (d_2+k)!/(d_2!\, k!)$ ([87]), and $m+1$ be equal to the number of linearly independent monomials of a degree $\leq d_1$ $(d_1 < d_2)$.

Let $\mathcal{X}$ be a bounded closed set in R^k having interior points.

Theorem 2.3. *Let $k \geq 2$. Then for the $\mathcal{X}$ and $\Psi_i(x)$ described above:*

1) If $N \geq n+1$ and F is a matrix of full rank, then any design ξ is an MB design

2) If $n+1 > N \geq M\left(\lfloor \frac{d_1+d_2}{2} \rfloor,\ k\right)$, then

a) a design ξ is an MB design if any polynomial of a degree $\leq d_2$ having zeros at the design points is orthogonal to all polynomials of a degree $\leq d_1$;

b) if an ideal formed by polynomials, having roots $x_1, \ldots, x_N$, has a homogeneous basis consisting of polynomials of a degree $\leq d_2$, then from the fact that the design is an MB design, it follows that a cubature formula with nodes at these design points, exact for all polynomials of a degree $\leq d_1 + d_2$, exist;

c) if an ideal (under the conditions of b)) has a homogeneous basis and at least one of the elements of this basis has a degree $q > d_2$ and is nonorthogonal to some polynomial of a degree $d_2 + d_1 - q$, then the indicated cubature formula does not exist.

3) If $N < M\left(\lfloor \frac{d_1+d_2}{2} \rfloor,\ k\right)$, MB designs do not exist.

Proof. 1),2), a) are proved similarly to the analogous hypothesis of Theorem 3.1.

Let us prove 2), b). Let $h_1(x), \dots, h_r(x)$ be an ideal homogeneous basis and the polynomials degrees $h_1(x), \dots, h_r(x) \le d_2$. Consider an arbitrary polynomial $P(x)$ of a degree $\le d_1 + d_2$ with roots at the design points. By definition of a homogeneous basis,

$$P(x) = \sum_{j=1}^{r} h_j(x)\, a_j(x), \tag{2.6}$$

where the degree of a polynomial a_j does not exceed the difference of polynomials $P(x)$ and $h_j(x)$ degrees and $a_j(x)$ vanishes at the design points.

By virtue of Sobolev's theorem, we are to prove that

$$\int P(x)\, \lambda(dx) = 0. \tag{2.7}$$

Let us represent $a_j(x)$ in the form

$$a_j(x) = \sum_i Q_{ij}(x)\, v_{ij}(x), \tag{2.8}$$

where $Q_{ij}(x)$ degree does not exceed d_2 minus degree of $h_j(x)$, the degree of $v_{ij}(x)$ does not exceed d_1. Then the degree of $h_j(x)\, Q_{ij}(x)$ is equal to n and this polynomial has roots at the design ξ points. By Remark 2.1 and by virtue of the fact that ξ is an MB design, polynomial $h_j(x)\, Q_{ij}(x)$ is orthogonal to all polynomials of a degree $\le d_1$. Taking into account (2.6) and (2.7), we have (2.8).

Hypothesis 2), c) is similarly proved.

Let us prove 3). According to Theorem 2.4 [87], at $N < M\left(\lfloor \frac{d_1+d_2}{2} \rfloor,\ k\right)$ points $x_1, \dots, x_N$ are the roots of a polynomial $Q(x)$ of a degree $\le \lfloor \frac{d_1+d_2}{2} \rfloor$. Let $Q_1(x)$ be a polynomial of a degree $d_2 - \lfloor \frac{d_1+d_2}{2} \rfloor$. Then $Q_1(x)\, Q(x)$ has roots $x_1, \dots, x_N$ and a degree $\le d_2$. If the design $\xi = \{x_1, \dots, x_N\}$ is an MB design, then, according to 2),a), polynomial $Q_1(x)\, Q(x)$ is orthogonal to all the polynomials of a degree $\le d_1$ and, hence, $Q(x)$ is orthogonal to all the polynomials of a degree $\le d_1 + d_2 - \lfloor \frac{d_1+d_2}{2} \rfloor$, which is impossible.

Theorem 2.3 is complete.

Consider an example of an MB design in nodes of which it is impossible to determine a cubature formula having exactness $d_1 + d_2$.

Let

$$k = 2,\ d_2 = 5,\ d_1 = 4,\ \lambda(dx) = dz_1\, dz_2,$$
$$x = (z_1, z_2),\ \mathcal{X} = \{(z_1, z_2);\ z_1, z_2 \ge 0,\ z_1 + z_2 \le 1\}$$

In the work [94], 19 points have been indicated which are the common zeros of the polynomials

$$f(z_1, z_2),\ \ g(z_1, z_2),\ \ h(z_1, z_2),\ \ \omega(z_1, z_2)$$

with the following properties: $f,\ g,\ h$ are polynomials of degree 5 orthogonal to all polynomials of a degree ≤ 4, ω is a polynomial of degree 6 such that $\int \omega(z_1, z_2)\, z_1^2\, z_2\, dz_1\, dz_2 \neq 0$. Here, $f,\ g,\ h,\ \omega$ is a homogeneous basis of an ideal of polynomials, having 19 roots indicated in [94]. Denote the set of these roots by ξ^*.

Let $P(z_1, z_2)$ be an arbitrary polynomial of a degree $\leq d_2 = 5$ having zeros at the design ξ^* points. According to the property of a homogeneous basis

$$P(z_1, z_2) = af + bg + ch,$$

where $a,\ b,\ c$ are constants. Hence,

$$\int P(z_1, z_2)\, \Psi_i(z_1, z_2)\, dz_1 dz_2 = 0, \quad i = 1, 2, \ldots, m,$$

since $f,\ g,\ h$ are orthogonal to all polynomials of a degree $\leq d_1 = 4$. Hence, by Remark 2.1, ξ^* is an MB design. Moreover, according to 2), c) it is impossible to determine from Theorem 2.3 a cubature formula which is exact for polynomials of a degree $\leq d_1 + d_2 = 9$ with nodes at points of ξ^*.

2.4. An iterative technique for determination of MB designs.

Let $\mathcal{X}$ be a compact metric space with a metric ρ and let $\Psi_1(x), \ldots, \Psi_n(x)$ be continuous functions. Denote a set of designs of dimension N such that $\rho(x_i, x_j) \geq (1 - \delta_{ij})\delta$ by $\Xi_{N,\delta}$. As is seen from Section 2.3, MB designs may not exist for some N. If an MB design exists, then it belongs to $\Xi_{N,\delta}$ for some $\delta > 0$.

Let us denote $\xi_s = \{x_1, \ldots, x_s\}$, and the points $x_1, \ldots, x_s$ will be considered to be distinct. The following statement is established by direct calculations (see [32]).

Lemma 2.7.

$$S(\xi_{s+1}) = S(\xi_s) + \frac{(I - S(\xi_s))\, f(x_{s+1})\, f^T(x_{s+1})(I - S(\xi_s))}{f^T(x_{s+1})\,(I - S(\xi_s))\, f(x_{s+1})}.$$

Denote $f_0(x) = f(x)$.

$$f_s(x) = (I - S(\xi_s))\, f(x); \quad \varphi_s(x) = T f_s(x), \quad s = 1, 2, \ldots, N.$$

Then, from Lemma 2.7, we have

$$f_{s+1}(x) = f_s(x) - \frac{f_s^T(x_{s+1})\, f_s(x)}{\|f_s(x_{s+1})\|^2}\, f_s(x_{s+1}), \tag{2.9}$$

$$\operatorname{tr} T S(\xi_{s+1}) T^T = \operatorname{tr} T S(\xi_s) T^T + \frac{\|\varphi_s(x_{s+1})\|^2}{\|f_s(x_{s+1})\|^2}, \tag{2.10}$$

at $x_{s+1} \neq x_1, \ldots, x_s$.

The determination of an MB design of dimension N is equivalent to the solution of the following problem

$$\operatorname{tr} TS(\xi)T^T \to \max_{\xi \in \Xi_{N,\delta}} .$$

Also, according to Theorem 2.1, a design ξ will be an MB design if

$$\operatorname{tr} TS(\xi)T^T = m.$$

Formula (2.10) gives a convenient representation for the value $g(x_1, \ldots, x_N) = \operatorname{tr} TS(\xi)T^T$ as the dependence of one of the coordinates x_i under others that are fixed (here each "coordinate" may be an element of R^k), which justifies the use of the coordinate descent technique.

The function

$$h_s(x) = \|\varphi_s(x)\|^2 / \|f_s(x)\|^2$$

is continuous over the compact set

$$V_s = \{x \in \mathcal{X};\ \rho(x, x') \geq \delta \ \text{ for all } \ x' \ \text{ such that } \ \|f_s(x')\| = 0\},$$

therefore $\max_{x \in V_s} h(x)$ exists.

Let us state an algorithm based on the cyclic coordinate descent technique.

Algorithm.

1. Let us assume $s = 0,\ l = 0,\ v = 0$;
2. Determine x_1: $h_0(x_1) = \max_{x \in \mathcal{X}} h_0(x)$;
3. Let $s = s + 1$.

Determine x_{s+1}: $h_s(x_{s+1}) = \max_{x \in \mathcal{X}} h_s(x)$;

4. If $s < N$, then go to 3, else $\xi^{(0)} = \{x_1, \ldots, x_N\}$;
5. Fix $\bar{x}_1 = x_2, \ldots, \bar{x}_{N-1} = x_N$ and determine $f_1(\bar{x}_1), \ldots, f_{N-1}(\bar{x}_{N-1})$ at points $\bar{x}_1, \ldots, \bar{x}_{N-1}$ and $\bar{x}_N$: $h_N(\bar{x}_N) = \max_{x \in V_N} h_N(x)$;
6. Let $l = l + 1$, $x_1 = \bar{x}_1, \ldots, x_N = \bar{x}_N$; if $l < N$, then go to 5;
7. Let $l = 0,\ v = v + 1,\ \xi^{(v)} = \{x_1, \ldots, x_N\}$;
8. If $\left|\operatorname{tr} TS(\xi^{(v)})T^T - \operatorname{tr} TS(\xi^{(v-1)})T^T\right| > \varepsilon$ where ε is a given exactness then go to 5.

Note that the counter s numerates the points at the initial design and the counter l numerates the points which are undergoing replacement: as soon as all N points are replaced, a comparison of the designs is executed (design $\xi^{(v)}$ is compared with design $\xi^{(v+1)}$ obtained by successively replacing all points). If designs $\xi^{(v)}$ and $\xi^{(v+1)}$ differ by the criteria $\operatorname{tr} TS(\xi)T^T$ more than in ε, then the successive replacing of points is executed again.

Since values $g_v = \operatorname{tr} TS(\xi^{(v)})T^T$ do not decrease, then the sequence g_v has the limit and there exists a subsequence $\xi^{(v_j)}$ converging to some limit $\xi^* \in \Xi_{N,\delta}$. As an objective function does not possess a convexity property in the general case, then investigation of the ξ^* characteristics represents a complicated problem which is beyond the scope of this book. In model examples with polynomials in one and two variables, the algorithm shows fast (2 or 3 iterations) convergence. In the general case, instead of $\xi^{(0)}$ defined in 1.- 4. of the algorithm, it is worthwhile randomly selecting $\xi^{(0)}$ and then repeatedly using 5.- 8. of the algorithm.

3. An approach to simulation data compression

Let there be a sequence of independent random variables $Z_1, Z_2, \dots, Z_n$ with the same distribution density $f(x, \Theta)$ given with an accuracy of a vector of unknown parameters $\Theta^T = (\theta_1, \dots, \theta_m)$. Let derivatives f'_{θ_i}, $i = 1, \dots, m$, exist and be continuous. Consider the case of $m = 1$. It is known that under some conditions [11, ch.3] for $n \to \infty$, the estimator of the maximal likelihood technique $\widehat{\Theta}$ is defined uniquely, $\widehat{\Theta} \to \Theta_u$, with probability equal to 1, where Θ_u is the true value of a parameter and $\widehat{\Theta}$ has an asymptotically normal distribution with expectation Θ_u and variance

$$\left[n \int \left[f'_\Theta(x, \Theta_u)\right]^2 / f(x, \Theta_u)\, dx\right].$$

Here this value cannot be less for any consistent estimator.

The values $Z_1, Z_2, \dots, Z_n$ can be interpreted as completion instants of an event in the system S, the behavior of which is characterized by the path of a random process $\{X_t\}$ simulated by means of a computer. In particular, if X_t is defined as the number of requests in a system of the $M/M/1$ class, then for $X_0 = 0$ the instance of the system returning into the 0 state has a distribution satisfying a certain equality [20].

The following question arises: if data transformation (the number of times that Z_i hits the segments $[0, x_1), [x_1, x_2), \dots, [x_{k-1}, x_k)$ is used instead of $Z_1, \dots, Z_n$) is performed to estimate parameter Θ in order to save computer memory and time taken to process results, then how much accuracy will be lost? The present chapter contains a series of results concerning the solution of this problem. These results was obtained in [82].

3.1. Formulation of the problem.

Let $Z_1, \dots, Z_n$ be independent random variables with a density $f(x, \Theta)$; $f(x, \Theta) = 0$ for $x < 0$ and is linearly dependent on the parameters $\Theta^T = (\theta_1, \dots, \theta_m)$:

$$f(x, \Theta) = \varphi_0(x) + \sum_{i=l}^{m} \theta_i \varphi_i(x), \quad \int_0^\infty \varphi_i(x)\, dx = 0, \quad \int_0^\infty \varphi_0(x)\, dx = 1,$$

where $\varphi_i(x)$ are known linearly-independent functions, given and continuous on $[0, \infty)$.

Let $\mathcal{X}_j$, $j = 1, 2, \dots, k+1 \geq m+1$ be mutually disjoint sets measurable by the Lebesgue measure such that $\cup \mathcal{X}_i = [0, \infty)$. Denote by $Y_j^{(1)}$ the number of points Z_i $(i = 1, 2, \dots, n_1)$ hit into $\mathcal{X}_j$, by $Y_j^{(2)}$ the same numbers for Z_i $(i = n_1 + 1, \dots, n)$, $Y_{(1)}^T = \left(Y_1^{(1)}, \dots, Y_k^{(1)}\right)$, $Y_{(2)}^T = \left(Y_1^{(2)}, \dots, Y_k^{(2)}\right)$.

Introduce the estimators

$$\widehat{\Theta}_{(1)} = (X^T X)^{-1} X^T \left(\alpha_1 Y_{(1)} - b_0\right),$$
$$\widehat{\Theta}_{(2)} = (X^T W^{-1}_{\widehat{\Theta}_{(1)}} X)^{-1} X^T W^{-1}_{\widehat{\Theta}_{(1)}} \left(\alpha_2 Y_{(2)} - b_0\right),$$

where $X = \left(\int_{\mathcal{X}_i} \varphi_j(x)\,dx\right)^{m,k}_{j,i=1}$, X^T has the order $m \times k$, $b_0 = \left(\int_{\mathcal{X}_i} \varphi_0(x)\,dx\right)^T$, $W_\Theta = E\left(YY^T \mid \Theta\right) - E\left(Y \mid \Theta\right)\left(EY(\Theta)\right)^T$, $Y = Y_{(2)}$, $\alpha_1 = 1/n_1$, $\alpha_2 = 1/(n - n_1)$.

Designate

$$D_{\widehat{\Theta}} = \left(\operatorname{Cov}\left(\hat{\theta}_i, \hat{\theta}_j\right)\right)^m_{i,j=1}.$$

Note that the observations $Y = Y_{(1)}$ $(Y = Y_{(2)})$ are described by the regression model

$$EY = \alpha^{-1}(X\Theta + b_0), \quad EYY^T - EY(EY)^T = \alpha^{-1} W_\Theta, \quad \alpha = \alpha_1 \quad (\alpha = \alpha_2).$$

Let the matrix $X^T X$ be nondegenerate. The problem of the investigation of estimators $\widehat{\Theta}_{(1)}$ and $\widehat{\Theta}_{(2)}$ and that of selecting optimal partitioning now appears.

3.2. The characteristics of the estimators.
The characteristics of the estimators $\widehat{\Theta}_{(1)}$ and $\widehat{\Theta}_{(2)}$ are similar to those of the estimators of the least-squares technique and the estimators of the weighted least-squares technique and are described by Theorem 3.1.

Theorem 3.1. *Under the conditions defined in Section 3.1:*
1) $\widehat{\Theta}_{(1)}$ *is an unbiased and consistent estimator of* Θ $(n_1 \to \infty)$*;*
2) $\widehat{\Theta}_{(2)}$ *is an asymptotically unbiased and consistent estimator of* Θ $(n - n_1 \to \infty)$*;*
3) $D_{\widehat{\Theta}_{(1)}} = \frac{1}{n_1}\left(X^T X\right)^{-1} X^T W_\Theta X (X^T X)^{-1}$*;*
4) $D_{\widehat{\Theta}_{(2)}} = \frac{1}{n-n_1}\left(X^T W^{-1}_{\widehat{\Theta}_{(1)}} X\right)^{-1} X^T W^{-1}_{\Theta} X \left(X^T W^{-1}_{\widehat{\Theta}_{(1)}} X\right)^{-1}$*;*
5) At n_1, $n - n_1 \to \infty$,

$$(n - n_1) D_{\widehat{\Theta}_{(2)}} X^T W^{-1}_\Theta X \to 1,$$

with probability equal to 1.

Proof. Since

$$E\widehat{\Theta}_{(1)} = E\left(X^T X\right)^{-1} X^T \left(\alpha Y_{(1)} - b_0\right) = \left(X^T X\right)^{-1} X^T X \Theta = \Theta,$$

then the estimator $\widehat{\Theta}_{(1)}$ is an unbiased estimator. Note that $\alpha = \frac{1}{n_1}$ and $\alpha Y^{(1)}_i = \int_{\mathcal{X}_i} dF_{n_1}(x)$, where $F_{n_1}(x)$ is an empirical distribution function. Designate

$$Y^* = (Y^*_1, \ldots, Y^*_k)^T, \quad Y^*_i = \int_{\mathcal{X}_i} dF(x).$$

We have $Y^* = X\Theta + b_0$. By virtue of the known theorem given in [11], $\alpha Y_i \to Y_i^*$ at $n_1 \to \infty$, with probability equal to 1. Therefore,

$$\widehat{\Theta}_{(1)} = \left(X^T X\right)^{-1} X^T \left(\alpha Y_{(1)} - b_0\right) \to \left(X^T X\right)^{-1} X^T X\Theta = \Theta,$$

with probability equal to 1 and, thus, $\widehat{\Theta}_{(1)}$ is a strongly consistent estimator of Θ. Statement 2) is proved similarly to statement 1) and assertions 3) – 5) are checked directly.

It is important for the considered problem that the matrix $X^T W_{\Theta_{(1)}}^{-1} X$ has a convenient form, i.e., Theorem 3.2 is valid.

Theorem 3.2. *Under the conditions of Theorem 3.1,*

$$(n - n_1) X^T W_\Theta^{-1} X = \left(\sum_{s=1}^{k+1} \frac{\int_{\mathcal{X}_s} \varphi_i(x)\, dx \int_{\mathcal{X}_s} \varphi_j(x)\, dx}{\int_{\mathcal{X}_s} f(x,\Theta)\, dx} \right)_{i,j=1}^{m}.$$

Proof. Note that

$$Y_i^{(2)} = \sum_{l=1}^{s} Y_i^{(2,l)}, \quad i = 1, 2, \dots, k,$$

where $s = n - n_1$, $Y_i^{(2,l)} = 1$ if $Z_l \in \mathcal{X}_i$, otherwise it is equal to 0. Hence,

$$(W_\Theta)_{i,j} = \sum_{l=1}^{s} \left(EY_i^{(2,l)} Y_j^{(2,l)} - EY_i^{(2,l)} EY_j^{(2,l)} \right) = s(A - aa^T),$$

where $A_{i,j} = \int_{\mathcal{X}_i \cap \mathcal{X}_j} f(x,\Theta)\, dx$, $a_i = \int_{\mathcal{X}_i} f(x,\Theta)\, dx$, $i,j = 1, \dots, k$, $A = (A_{i,j})$, $a = (a_i)$. Use the formula $(I + cd^T)^{-1} = I - cd^T(1 + d^T c)$ for $c = -a$, $d^T = a^T A^{-1}$ [38].

We have

$$sW_\Theta^{-1} = (A - aa^T)^{-1} = A^{-1}(I - aa^T A^{-1})^{-1} =$$
$$= A^{-1} + A^{-1} aa^T A^{-1} / (1 - a^T A^{-1} a).$$

Direct calculations give

$$1 - a^T A^{-1} a = 1 - \sum_{i=1}^{k} \int_{\mathcal{X}_i} f(x,\Theta)\, dx = \int_{\mathcal{X}_{k+1}} f(x,\Theta)\, dx,$$

$$X^T A^{-1} X = \left(\sum_{l=1}^{k} \int_{\mathcal{X}_l} \varphi_i(x)\, dx \int_{\mathcal{X}_l} \varphi_j(x)\, dx \Big/ \int_{\mathcal{X}_l} f(x,\Theta)\, dx \right)_{i,j=1}^{m}, \tag{3.1}$$

$$X^T A^{-1} aa^T A^{-1} X = \left(X^T A^{-1} a\right) \left(X^T A^{-1} a\right)^T = vv^T,$$

where

$$v = \left(\int_0^\infty \varphi_i(x)\,dx - \int_{\mathcal{X}_{k+1}} \varphi_i(x)\,dx\right)_{i=1}^m = -\left(\int_{\mathcal{X}_{k+1}} \varphi_i(x)\,dx\right)_{i=1}^m .$$

Finally, we have

$$sX^T W_\Theta^{-1} X = X^T A^{-1} X + X^T A^{-1} a a^T A X/(1 - a^T A^{-1} a) =$$

$$= \left(\sum_{l=1}^{k+1} \frac{\int_{\mathcal{X}_l} \varphi_i(x)\,dx \int_{\mathcal{X}_l} \varphi_j(x) dx}{\int_{\mathcal{X}_l} f(x,\Theta)}\,dx\right)_{i,j=1}^m . \tag{3.2}$$

The proof is complete.

Designate $M_R = X^T W_\theta^{-1} X$ (index R means that M_R refers to the estimators of the regression model parameters). A relation with the Fisher information matrix

$$M = \left(\int \frac{\varphi_i(x)\,\varphi_j(x)}{f(x,\Theta)}\,dx\right)_{i,j=1}^m$$

appears at the refinement of the partitioning $\mathcal{X}_1, \ldots, \mathcal{X}_{k+1}$. Let $\mathcal{X}_1^{(l)}, \ldots, \mathcal{X}_{k_l+1}^{(l)}$, $l = 1, 2, \ldots$ be such that

$$\max_{1 \le l \le k_1+1} \mu\left(\mathcal{X}_j^{(l)}\right) \to 0 \quad (l \to \infty), \tag{$*$}$$

where μ is the Lebesgue measure.

Denote

$$X_l = \left(\int_{\mathcal{X}_i^{(l)}} \varphi_j(x)\,dx\right), \quad M_R^{(l)} = X_l^T W_\Theta^{-1} X_l.$$

Theorem 3.3. *If the conditions of Theorem 3.1 and* $(*)$ *are valid, then*

$$\left(M_R^{(l)}\right)_{ij} \to (M)_{ij}, \quad i,j = 1, \ldots, m, \quad l \to \infty.$$

Proof. The proof of Theorem 3.3 is presented for the case of the function

$$g(x) = \varphi_i(x)\varphi_j(x) \,/\, f(x,\Theta)$$

is integrable in the Riemann sense, and $\mathcal{X}_j^{(l)}$ has the form of $\left[x_{j-1}^{(l)}, x_j^{(l)}\right]$, $x_0^{(l)} = 0$, $x_{k_1+1}^{(l)} = \infty$. Under these suppositions, Darboux's sums

$$\sum_{j=1}^{k_l+1} \bar{g}_{j,l}\,\mu\left(\mathcal{X}_j^{(l)}\right), \quad \sum_{j=1}^{k_l+1} \underline{g}_{j,l}\,\mu\left(\mathcal{X}_j^{(l)}\right),$$

where $\bar{g}_{j,l} = \sup_{x\in\mathcal{X}_j^{(l)}} g(x)$, $\underline{g}_{j,l} = \inf_{x\in\mathcal{X}_j^{(l)}} g(x)$, tend to limit $\int_0^\infty g(x)\,dx = (M)_{i,j}$. Since

$$\sum_s \underline{g}_{s,l}\mu\left(\mathcal{X}_s^{(l)}\right) \le \left(M_R^{(l)}\right)_{i,j} \le \sum_s \bar{g}_{s,l}\mu\left(\mathcal{X}_s^{(l)}\right),$$

hence the conclusion of the theorem follows.

The results described above give grounds for the investigation of partitioning $\{\mathcal{X}_l\}$ characteristics.

3.3. The experimental designs.

The choice of an optimal in some sense collection $\{\mathcal{X}_j\}$ is similar to the problems of the regression experimental design [31].

Let us put forward the following definition.

Definition 3.1. A collection of measurable sets $\mathcal{X}_i$ satisfying the conditions $\mathcal{X}_i \cap \mathcal{X}_j = 0$ $(i \neq j)$, $\cup\mathcal{X}_i = [0,\infty)$, $i,j = 1,2,\dots,k+1$ is called an experimental design (in the problem of data compression).

Further, we will restrict ourselves to the consideration of experimental designs of the type

$$\mathcal{X}_i = \kappa_i = [x_{i-1}, x_i), \quad x_0 = 0, \quad x_{k+1} = \infty, \tag{**}$$

which are very easy to realize.

Definition 3.2. An experimental design (in the problem of data compression) of the type (**) is called simple.

Let us consider the D-criterion analogue as an optimum criterion [31]

$$\det X^T W_\Theta^{-1} X \to \sup, \tag{***}$$

where Θ is a fixed value of the vector of the parameter (obtained in the form of the estimator $\widehat{\Theta}_{(l)}$), and the supremum will be considered over a set of simple designs with a fixed $(k+1)$ number of sets k_j. As is known, this criterion leads to the minimization of the estimator Θ's ellipsoid volume concentration, if Y has a normal distribution (which is valid in this case in the asymptotically sense).

For $k+1 = m+1$, the following result takes place.

Theorem 3.4. *If $k = m$, then*

$$\det X^T M_\Theta^{-1} X = \det^2 (\Phi_i(x_j)) \Big/ \prod_{l=1}^{m+1} \int_{\kappa_l} f(x,\Theta)\,dx\ ,$$

where $\Phi_l(x) = \int_0^x \varphi_l(t)\,dt$, $\Phi_0 \equiv 1$, $i,j \in 0:m$.

Proof. Note that

$$\left(\sum_{s=1}^m \int_{\kappa_s} \varphi_i(x)\,dx \int_{\kappa_s} \varphi_j(x)\,dx \Big/ \int_{\kappa_s} f(x,\Theta)\,dx\right)_{i,j=1}^m = B^T B,$$

where

$$B = \left(\left(\int_{\kappa_l} \varphi_i(x)\,dx\right) \Big/ \left(\int_{\kappa_l} f(x,\Theta)\,dx\right)^{1/2}\right)_{i,l=1}^{m}.$$

Further,

$$\det B = \left(\prod_{l=1}^{m} \int_{\kappa_l} f(x,\Theta)\,dx\right)^{-1/2} \det\left(\int_{\kappa_i} \varphi_j(x)\,dx\right)_{i,j=1}^{m}.$$

But

$$\det \begin{pmatrix} 1 & \dots & 1 \\ \Phi_1(x_0) & \dots & \Phi_1(x_m) \\ \vdots & & \vdots \\ \Phi_m(x_0) & \dots & \Phi_m(x_m) \end{pmatrix} =$$

$$= \det \begin{pmatrix} \Phi_1(x_1) - \Phi_1(x_0) & \dots & \Phi_1(x_m) - \Phi_0(x_{m-1}) \\ \vdots & & \vdots \\ \Phi_m(x_1) - \Phi_m(x_0) & \dots & \Phi_m(x_m) - \Phi_m(x_{m-1}) \end{pmatrix}$$

and

$$\Phi_i(x_{j+1}) - \Phi_i(x_j) = \int_{x_j}^{x_{j+1}} \varphi_i(x)\,dx = \int_{\kappa_{j+1}} \varphi_i(x)\,dx.$$

Hence, by virtue of (3.1), (3.2) and the formula $\det(C + VV^T) = \det C(1 + V^T C^{-1} V)$, the proof of the theorem is complete.

Thus, by virtue of Theorem 3.4 and known results [31, 38], points $0, x_1, \dots, x_k$ form the so-called exact experimental design to evaluate parameters $\Theta_0, \Theta_l, \dots, \Theta_m$ and a regression of the type

$$Y(x_j) = \sum_{i=1}^{m} \Theta_i \Phi_i(x_j) + \Theta_0 + \varepsilon_j,$$

where $E\varepsilon_j = 0$, $E\varepsilon_l\varepsilon_j = \delta_{lj} \int_{k_{l+1}} f(x,\Theta)\,dx$, $l, j = 0, 1, \dots, k$.

It enables us to use known analytical results and numerical techniques of the design theory of regression experiments [31, 38].

For an arbitrary k to receive the designs, optimal in the sense of the criterion $(***)$, we can numerically use the cyclic coordinate-wise descent technique. Such a method is justified by the possibility of a direct calculation of the derivative of the function $h(x_1, \dots, x_k) = \det X^T W_\Theta^{-1} X$ with respect to one of the coordinates $x_1, \dots, x_k$ under others which are fixed. A direct calculation gives us for the case $m = 1$

$$h'_{x_i} = \psi(x_{i-1}, x_i) - \psi(x_i, x_{i+1}),$$

where

$$\psi(x,y) = \left\{ \frac{\left(\int_x^y \varphi_1(t)\,dt\right)^2}{\int_x^y f(t,\Theta)dt} \right\}' = \{2\,(\Phi_1(y) - \Phi_1(x))\,\varphi_1(y)(F(y,\Theta)-$$

$$- F(x,\Theta)) - f(x,\Theta)(\Phi_1(y) - \Phi_1(x))^2\} \Big/ [F(y,\Theta) - F(x,\Theta)]^2\,,$$

$$F(x,\Theta) = \int_0^x f(t,\Theta)\,dt.$$

In the general case, similar formulas can be obtained on the basis of the known presentation [31]

$$\det(A + bb^T) = \det A\,\left(1 + b^T A^{-1} b\right)$$

if we suppose that

$$A = \left(\sum_{s \neq i, i+1} \frac{\int_{K_s} \varphi_j(x)\,dx \int_{K_s} \varphi_q(x)\,dx}{\int_{K_s} f(x,\Theta)\,dx} \right),$$

$$b_{i+1} = \left(\int_{K_{i+1}} \varphi_j(x)\,dx \Big/ \left| \int_{K_{i+1}} f(x,\Theta)\,dx \right|^{1/2} \right), \qquad i = 0, 1,$$

and calculate sequentially $\det M_R = \det\left(A + b_1 b_1^T + b_2 b_2^T\right)$.

The construction of optimal designs for concrete functions enables us to solve the problem stated above. Let us give a simple example of this solution.

Example. Let

$$f(x,\Theta) = \lambda\Theta e^{-\lambda x} + \mu(1-\Theta)e^{-\mu x}, \quad \mu > \lambda,$$

and let the true value of a parameter Θ be equal to 1.

Direct calculation gives

$$\det M = M = \int_0^\infty \frac{1}{\lambda}\,\left(\lambda e^{-\lambda x} - \mu e^{-\mu x}\right)^2 e^{\lambda x}\,dx = (\mu - \lambda)^2 \,/\, (\lambda(2\mu - \lambda)).$$

For $k = 1$, we have

$$\det M_R = M_R = \frac{\left(\int_0^{x_1} \left(\lambda e^{-\lambda x} - \mu e^{-\mu x}\right)\,dx\right)^2}{\left(1 - e^{-\lambda x_1}\right) e^{-\lambda x_1}}.$$

Note that the simulation time decreases with data compression. In the usual approach, the mean time of one-path simulation is proportional to the expectation of a random ν

with the density $\lambda e^{-\lambda x}$, i.e. $1/\lambda$, and under data compression, the mean simulation time is proportional to $E\nu_1$, where $\nu_1 = \min\{\nu, x_1\}$ and is equal to $\left(1 - e^{-\lambda x_1}\right)/\lambda$.

Therefore, it is necessary to compare the values λM and $\lambda M_R / \left(1 - e^{-\lambda x_1}\right)$. We have

$$\lambda M = (\mu - \lambda)^2 / (2\mu - \lambda),$$

$$\lambda M_R \Big/ \left(1 - e^{-\lambda x_1}\right) = \lambda \left(\frac{e^{-\lambda x_1} - e^{\mu x_1}}{1 - e^{-\lambda x_1}}\right)^2 e^{-\lambda x_1},$$

e.g., for $x_1 = \ln((2\mu - \lambda)/\lambda) \,/\, (\mu - \lambda)$ and $\mu = 2\lambda$ we have

$$n\mathcal{D}_{\hat{\Theta}} ET_{\mathrm{sim}} \sim \left(\frac{1}{3}\lambda\right)^{-1},$$

for the usual approach (T_{sim} is the simulation time), and for estimators $\widehat{\Theta}_{(2)}$ with data compression

$$(n - n_1)\mathcal{D}_{\hat{\Theta}(2)} ET_{\mathrm{sim}} \sim \left(\frac{1}{3}\lambda\right)^{-1}.$$

Note that the condition of Z_i nonnegativity does not matter in Theorems 3.1–3.3.

4. The duality theorem and E-optimality

Let $y(x)$ be the result of an experiment at a point x, satisfying the standard regression equation

$$Ey(x) = f^T(x)\theta, \quad x \in \mathcal{X},$$

where E is the symbol of expectation, $f^T(x) = (f_1(x), \ldots, f_m(x))$ is a vector of functions, continuous at the bounded closed set $\mathcal{X}$, which characterizes the experiment performance conditions; $\theta^T = (\theta_1, \ldots, \theta_m)$ is a vector of the estimated parameters. The aim of the experiment is to evaluate θ (functions $f_i(x)$ are supposed to be known). As for the errors of the experiment, let us assume that they are uncorrelated and have equal variances. We will consider so-called approximate designs consisting of points and observation weights

$$\xi = \{x_1, \ldots, x_n;\ \mu_1, \ldots, \mu_n\}, \quad x_i \in \mathcal{X},$$

where

$$\mu_i > 0, \quad \sum \mu_i = 1,$$

which are probability measures concentrated at a finite number of points.

It is known that a variance-covariance matrix of estimators θ by the least-squares technique has the form [31]

$$\sigma^2 M^{-1}, \quad M = \sum_{i=1}^{n} f(x_i)\, f^T(x_i)\, \mu_i,$$

where M is an information matrix; σ^2 is the variance of $y(x)$.

A design ξ^* for which [31]

$$\lambda_{\min}(M(\xi^*)) = \sup \lambda_{\min}(M(\xi)),$$

where $\lambda_{\min}(M)$ is the minimal eigenvalue of the matrix M and ξ runs through the set of all approximate experimental designs, is called an E-optimal experimental design.

The E-optimum criterion should be used in cases when the investigator is interested in various linear combinations of the parameters being estimated. If the sum of coefficient squares in the linear combination is fixed, then the E-optimum provides the maximum possible accuracy of evaluating a least-favourable situation.

Let us consider the problem: find

$$\min 1/\operatorname{tr} A, \tag{4.1}$$

where A runs through a set of all nonnegatively defined $m \times m$ matrices satisfying the condition

$$\max_{x\in\mathcal{X}} f^T(x) A f(x) \leq 1. \tag{4.2}$$

Theorem 4.1. *The problem (4.1)–(4.2) and the problem of finding E-optimum designs are connected by the duality relation*

$$\min 1/\operatorname{tr} A = \sup_{\xi} \lambda_{\min}(M(\xi)).$$

Proof. Introduce the notations to be used further. Let $\mathcal{U}$ be the class of nonnegatively defined $m \times m$ matrices A such that $\operatorname{tr} A = 1$ and let $\mathcal{M}$ be the class of information matrices corresponding to all approximate experimental designs

$$g(M) = \inf_{A\in\mathcal{U}} \operatorname{tr} AM,$$

$$\varphi(A) = \sup_{M\in\mathcal{M}} \operatorname{tr} AM.$$

The set $\mathcal{M}$ is convex and compact [47]. Evidently, a set $\mathcal{U}$ is convex and compact. The function $\operatorname{tr} AM$ is simultaneously convex and concave with respect to any argument (this is the corollary of the main properties of the operation tr).

Let us go over to the proof of Theorem 4.1.

Lemma 4.1. *The function* tr AM *has at least one saddle point for $A \in \mathcal{U}$, $M \in \mathcal{M}$, while for any saddle point (A^*, M^*), the equalities*

$$\varphi(A^*) = \inf_{A\in\mathcal{U}} \sup_{m\in\mathcal{M}} \operatorname{tr} AM = g(M^*) = \sup_{m\in\mathcal{M}} \inf_{a\in\mathcal{U}} \operatorname{tr} AM \tag{4.3}$$

are valid.

Note, that since the operation tr is continuous, the sets $\mathcal{U}$ and $\mathcal{M}$ are compact, we can write min and max instead of inf and sup everywhere.

Lemma 4.2. *For any nonnegatively defined matrix M,*

$$\min_{a\in\mathcal{U}} \text{tr } AM = \lambda_{\min}(M).$$

Lemma 4.1 is the statement of the Neumann minimax theorem [44] for the present case.

The proof of Lemma 4.2. Let P be a proper subspace corresponding to $\lambda_{\min}(M)$. Then [8]

$$\lambda_{\min}(M) = p^T M p < q^T M q,$$

where $p^T p = 1,\ \ p \in P,\ \ q^T q = 1,\ \ q \notin P$.

By Caratheodory's theorem, $A \in \mathcal{U}$ is represented as

$$A = \sum_{i=1}^{l} q_i q_i^T \mu_i,$$

where $\sum \mu_i = 1$, $\mu_i > 0$, $l \le m(m+1)/2$. Now we have

$$\text{tr } AM = \sum_{i=1}^{l} q_i^T M q_i \mu_i \ge \lambda_{\min}(M)$$

where the equality is attained at $q_i \in P\ \ (i = 1, 2, \dots, l)$.

The proof of Lemma 4.2 is complete.

According to Lemma 4.2, the latter expression in relation (4.3) has the form

$$\max_{M\in\mathcal{M}} \min_{A\in\mathcal{U}^*} \text{tr } AM = \max_{M\in\mathcal{M}} \lambda_{\min}(M). \tag{4.4}$$

Show that

$$\min_{A\in\mathcal{U}} \max_{M\in\mathcal{M}} \text{tr } AM = \min_{A\in\mathcal{U}^*} 1/\text{tr } A, \tag{4.5}$$

where $\mathcal{U}^*$ is the set of nonnegatively defined matrices satisfying (4.2).

In fact, let

$$A^* = arg \min_{A\in\mathcal{U}} \max_{M\in\mathcal{M}} \text{tr } AM$$

(the existence of A^* follows from Lemma 4.1).

We have

$$\max_{M\in\mathcal{M}} \text{tr } A^* M/\lambda_{\min} = \max_{x\in\mathcal{X}} f^T(x)\,(A^*/\lambda_{\min})\,f(x) = 1,$$

since

$$\min_{A\in\mathcal{U}} \max_{M\in\mathcal{M}} \text{tr } AM = \lambda_{\min} \tag{4.6}$$

in accordance with (4.3) and (4.4) (here $\lambda_{\min} = \max_{M\in\mathcal{M}} \lambda_{\min}(M)$).

Hence, $A^*/\lambda_{\min} \in \mathcal{U}^*$.
Further, tr $A^* = 1$ (since $A^* \in \mathcal{U}$). Thus

$$\min_{A \in \mathcal{U}^*} 1/\operatorname{tr} A \leq \lambda_{\min}/\operatorname{tr} A^* = \lambda_{\min}. \tag{4.7}$$

Show that

$$\min_{A \in \mathcal{U}^*} 1/\operatorname{tr} A^* \geq \lambda_{\min}. \tag{4.8}$$

Conversely, let there exists $A_0 \in \mathcal{U}^*$ such that $\lambda_{\min} \operatorname{tr} A_0 > 1$.
Then, since $A_0/h \in \mathcal{U}^*$ where $h = \operatorname{tr} A_0$,

$$\max_{M \in \mathcal{M}} \operatorname{tr} A_0 M = h \max_{M \in \mathcal{M}} \operatorname{tr} (A_0/h) M \geq h \min_{A \in \mathcal{U}} \max_{M \in \mathcal{M}} \operatorname{tr} AM = h\lambda_{\min} > 1,$$

which is impossible, since $A_0 \in \mathcal{U}^*$ and, therefore, satisfies condition (4.2), and

$$\max_{M \in \mathcal{M}} \operatorname{tr} A_0 M = \max_{X \in \mathcal{X}} f^T(x) A_0 f(x).$$

Now, relation (4.5) follows from (4.6), (4.7), (4.8). Comparing (4.4), (4.5), (4.6), we have

$$\min_{A \in \mathcal{U}^*} 1/\operatorname{tr} A = \max_{M \in \mathcal{M}} \lambda_{\min}(M). \tag{4.9}$$

Thus, Theorem 4.1 is complete.

The following characteristics of the solution of problem (4.1)–(4.2) can be obtained as a corollary of Theorem 4.1.

Let $\{\xi_\alpha\}$ be a set of all E-optimal designs and P_α be a space corresponding to $\lambda = \lambda_{\min}(M(\xi_\alpha))$ (its dimension from 1 to m). Denote $P = \cap P_\alpha$.

Corollary 4.1. *P is nonempty and any solution of problem (4.1)–(4.2) has the form*

$$A^* = \lambda_{\min}^{-1} \sum_{i=1}^{s} p_i p_i^T \lambda_i,$$

where s is dimension of P and $\{p_i\}$ is an orthonormalized basis of P, $\lambda_i \geq 0$, $\sum \lambda_i = 1$.

Proof. Since $A^* \geq 0$, then A^* can be represented in the form

$$A^* = h \sum_{i=1}^{m} q_i q_i^T \lambda_i,$$

where $h = \operatorname{tr} A^*$, $\lambda_i \geq 0$, $\sum \lambda_i = 1$, $\{q_i\}_{i=1}^m$ is an orthonormalized basis of R^m.

According to (4.9),

$$1/\operatorname{tr} A^* = \lambda_{\min}$$

and, therefore, $h = \text{tr } A^* = 1/\lambda_{\min}$. By Lemma 4.2,

$$\text{tr } A^* M^*/h = \lambda_{\min},$$

where $M^* = M(\xi^*)$, ξ^* is E-optimal design. Besides,

$$q_i^T M^* q_i > \lambda_{\min} \tag{4.10}$$

for some M^* if $q_i \notin P = \cap P_\alpha$. So we have

$$\text{tr } A^* M^*/h = \sum_{i=1}^{m} \lambda_i q_i^T M^* q_i = \lambda_{\min}.$$

This equality implies that the coefficients correspondings to $q_i \notin P$ in a representation for A^* should be zero. So the matrix A^* has the required form and, hence, Corollary 4.1 is complete.

The solution of problem (4.1)–(4.2) has a particularly simple form in the special case when there exists an E-optimal design ξ such that multiplicity of $\lambda_{\min}(M(\xi))$ is equal to 1. In this case, the solution of (4.1)–(4.2) has the form $A^* = \lambda^{-1} pp^T$, where p is an eigenvector corresponding to $\lambda = \lambda_{\min}(M(\xi))$.

Also, from Theorem 4.1, a useful necessary condition of E-optimality follows. Denote an E-optimal design by

$$\xi^* = \{x_1^*, \dots, x_n^*;\ \mu_1^*, \dots, \mu_n^*\}.$$

Corollary 4.2. *At the points of any E-optimal design ξ^*, we have*

$$f^T(x_i^*) A^* f(x_i^*) \equiv 1 \quad (i = 1, 2, \dots, n),$$

where A^* is a solution of problem (4.1)-(4.2).

Proof. Let A^* be the solution of (4.1)–(4.2), then, as in the proof of Corollary 4.2, we have

$$1 = \max_{M \in \mathcal{M}} \text{tr } A^* M = \text{tr } A^* M^*.$$

Hence,

$$1 = \max_{X \in \mathcal{X}} f^T(x) A^* f(x) = \text{tr } A^* M^*.$$

Therefore, if $\xi^* = \{x_1^*, \dots, x_n^*;\ \mu_i^*, \dots, \mu_n^*\}$ is an E-optimal design, then

$$f^T(x_i^*) A^* f(x_i^*) \geq 1.$$

Multiplying both sides of this inequality by μ_1^* and putting them together, we have

$$\text{tr } A^* M(\xi^*) \geq 1$$

and

$$\text{tr } A^* M(\xi^*) = 1,$$

if and only if

$$f^T(x_i^*) A^* f(x_i^*) = 1 \quad (i = 1, 2, \dots, n),$$

i.e. Corollary 4.2 is valid.

Corollary 4.3. *The design ξ is an E-optimal design if and only if there exists a matrix A such that*
1) $\max_{X\in\mathcal{X}} f^T(x)\,A\,f(x) = 1$,
2) $\operatorname{tr} A = 1/\lambda_{\min}(M(\xi))$.

In fact, the proof of Corollary 4.3 is contained in the proof of Theorem 4.1.
Let $f^T(x) = (1, \sin x, \cos, \dots, \sin kx, \cos kx)$, $\mathcal{X} = [0, 2\pi]$.

Theorem 4.2. *Designs ξ, for which $M(\xi) = \operatorname{diag}\|1, 1/2, \dots, 1/2\|$, are E-optimal.*

Proof. Let $A = \operatorname{diag}\|0, 1/k, \dots, 1/k\|$, dimension of $A - (2k+1)\times(2k+1)$.
We have $\operatorname{tr} A = 2 = 1/\lambda_{\min}(M(\xi))$, $f^T(x)\,A\,F(x) = (1/k)\{\sin^2 x + \dots + \sin^2 kx + \cos^2 kx\} \equiv 1$. By Corollary 4.3 a design ξ is E-optimal. The theorem is complete.

A set of designs satisfying the conditions of Theorem 4.2 is described in [39]. These designs are also D-optimal. In particular, the design

$$\xi = \left\{x_1, \dots, x_{2k+1};\ \frac{1}{2k+1} \cdots \frac{1}{2k+1}\right\},$$

$x_i = 2\pi(i-1)/2k$. is a design of that kind.

Results of this section was published in [82]. Similar results by the more complicated technique was obtained in [92]. The development of results of this section for polynomial regression on arbitrary segments can be found in [85].

5. Optimal experimental design for an exponential regression

This section is devoted to optimal experimental designs for the case of observation results satisfy the relations

$$y_j = \sum_{i=1}^{k} \sum_{l=0}^{i_k} a_{il} x_j^l \exp(-\lambda_i x_j) + \varepsilon_j \quad x_j \in [0, +\infty), \tag{5.1}$$

$j = 1, 2, \dots, N$, where $a_{10}, \dots, a_{1i_1}, \dots, a_{k0}, \dots, a_{ki_k}$, $\lambda_1, \dots, \lambda_k$ are estimated parameters and ε_j are random errors:

$$E\,\varepsilon_j = 0, \quad E\varepsilon_j\varepsilon_s = 0 \ \ (j \neq s), \quad E\varepsilon_j^2 = \sigma^2 > 0.$$

These models appear because of the approximation of distribution density, in particular, in queueing theory problems. Moreover, an important class of solutions for systems of linear differential equations has a type of regression function in the right side of (5.1), therefore such models are widely spread in experimental practice. Specifically, the splitting processes in nuclear physics and optics are described with the help of them [104].

To shorten the representation, introduce a formally simpler model

$$Y_j = \sum_{i=1}^{k} a_i \exp(-\lambda_i x_j) + \varepsilon_j, \tag{5.2}$$

To shorten the representation, introduce a formally simpler model

$$Y_j = \sum_{i=1}^{k} a_i \exp(-\lambda_i x_j) + \varepsilon_j, \tag{5.2}$$

where $j = 1, 2, \dots, N$. Evidently all the reasoning and results are extended to models of the type (5.1).

In this section, we will consider two widespread conceptions of experimental design. A collection of unnecessary distinct points $\tau = (x_1, \dots, x_N)$, $x_i \in [0, +\infty)$, $i = 1, 2, \dots, N$, will be called an exact design. An approximate experimental design is the collection of the points and weights of observations: $\zeta = \{x_1, \dots, x_n;\ p_1, \dots, p_n\}$, where n is arbitrary, $p_i \geq 0$, $\sum_1^n p_i = 1$. Let us consider the information matrix determinant as a criterion of the optimum design.

The main difficulty of this problem is the fact that the information matrix determinant depends on the parameters under estimation $(\lambda_1, \dots, \lambda_k)$. In this connection, we will study two types of designs. Designs maximizing the minimum of a certain set of information matrix determinants over $\lambda_1, \dots, \lambda_k$ in the class of all approximate (exact) designs are called D-minimax (exact D-minimax) designs. Designs maximizing the information matrix determinant in a class of all approximate (in the class of all approximates concentrated at the least possible number of points) designs under a fixed value of parameters $\lambda_1, \dots, \lambda_k$ are called locally D-optimal (saturated) designs.

The investigation of locally D-optimal designs in the general case of parametres which are nonlinear over models was originated in the work [19]. In this section, it is shown that in the case of the exponential regression in (5.1), the problem of finding D-optimal designs is reduced to one of finding locally D-optimal designs (under natural conditions). The properties of saturated locally D-optimal designs as vector functions of the parameters under estimation are also studied. This section is based on the paper [78].

5.1. D-minimax designs.

Let $\tau = (x_1, \dots, x_N)$ be an exact experimental design, i.e. a collection of unnecessary distinct points $x_1, \dots, x_N$ from $\mathcal{X} = [0, -\infty)$. The information matrix determinant for the regression function (5.2) after factorizing by the Binet-Cauchy formula is presented in the form

$$\det M(\tau, A, \Lambda) = a_1^2 \dots a_k \sum_{1 \leq \alpha_1 < \dots < \alpha_{2k} \leq N} \Big[\det \| \exp(-\lambda_1 x_{\alpha_i}), -x_{\alpha_i} \exp(-\lambda_i x_{\alpha_i}), \dots$$
$$\dots, \exp(-\lambda_k x_{\alpha_i}), -x_{\alpha_i} \exp(-\lambda_k x_{\alpha_i}) \|_{i=1}^{2k} \Big]^2, \quad N \geq 2k \tag{5.3}$$

(for the case of $N < 2k$ $\det M(\tau, A, \Lambda) \equiv 0$); $a = (a_1, \dots, a_k)^T$, $\Lambda = (\lambda_1, \dots, \lambda_k)^T$.

Let $a_i \neq 0$, $i = 1, \dots, k$. Evidently, the point, for which the maximum of $\det M(\tau, A, \Lambda)$ over τ is attained, is independent of A. For approximate designs (ξ) we can write a formula similar to (5.3), therefore the value ξ^*, for which the maximum of the information matrix determinant in the class of approximate designs is attained, is also independent of A.

Here and henceforth, we will consider that x_i $(i = 1, 2, \dots, N)$ and $-\lambda_j$ $(j = 1, 2, \dots, k)$ are labelled in ascending order: $x_1 \le x_2 \le \dots \le x_N$, $-\lambda_1 \le -\lambda_2 \le \dots \le -\lambda_k$. As for the vector Λ, let us suppose that $\Lambda \in \Omega$, where Ω is a set of a-priori information. Moreover, Ω has the form

$$\Omega = \{\Lambda \in R^k;\ \ \lambda_1 \le d,\ \lambda_i - \lambda_{i+1} \ge \rho_i,\ \ (i = 1, 2, \dots, k-1)\}, \tag{5.4}$$

where d, ρ_i $(i = 1, 2, \dots, k-1)$ are arbitrary fixed positive numbers $\sum \rho_i < d$.

The statement $\Lambda \in \Omega$ can be considered as a natural complement for equation (5.1), since $\det M(\tau, A, \Lambda) \equiv 0$ if Λ is such that $\lambda_i = \lambda_{i+1}$ for some i. If the number of distinct points (N') of the design τ is less than the number of parameters under estimation $(2k)$, then the information matrix determinant is evidently equal to zero. Let $N' \ge 2k$.

Theorem 5.1. *For any fixed design* $\tau = \{x_1, \dots, x_s\}$, *the number of distinct points which is not less than* $2k$, *the minimum of the value* $\det M(\tau, A, \Lambda)$ *over* $\Lambda \in \Omega$ *is attained if and only if* $\Lambda = \Lambda^* = \left[d,\ d - \rho_1, \dots, d - \sum_1^{n-1} \rho_i\right]^T$.

Proof. First consider an experimental design of the type $\tau' = \{x_1, \dots, x_{2k}\}$, $0 \le x_1 < \dots < x_{2k}$. Only one summand remains in the right side of (5.3), and $\det M(\tau, A, \Lambda) = a_1^2 \dots a_k^2\, [\bar{M}(\tau', \Lambda)]^2$, where

$$\bar{M}(\tau', \Lambda) = (-1)^k \det \left\| e^{-\lambda_1 x_s}, -x_s e^{-\lambda_1 x_s}, \dots, e^{-\lambda_k x_s}, -x_s e^{-\lambda_k x_s} \right\|_1^{2k}. \tag{5.5}$$

Further, referring to the Tchebyshev property of an exponential system, we will bear in mind that the function

$$\sum_{i=1}^{k} \sum_{j=0}^{t_i} a_{ij}\, x^j \exp(\lambda_i x)$$

under arbitrary fixed real $\{a_{ij}\}$, $\{\lambda_i\}$, $\sum a_{ij}^2 > 0$ vanishes not more than $n = \left(\sum_{i=1}^k t_i\right) + (k-1)$ times on account of multiplicity and that the determinant

$$\det \left\| e^{\lambda_1 x_s}, x_s e^{\lambda_1 x_s}, \dots, x_s^{t_k} e^{\lambda_k x_s} \right\|_{s+1}^{n+1}$$

is strictly positive at any $x_1 < x_2 < \dots < x_{n+1}$, $\lambda_1 < \dots < \lambda_k$. The proof of this fact is a simple modification of the reasoning, presented in [47].

According to the Tchebyshev property of an exponential system, it follows in particular that the function $\bar{M}(\tau', \Lambda)$ defined in (5.5) is strictly positive at $\Lambda \in \Omega$, $0 \le x_1 < \dots < x_{2k}$. Hence, to prove the theorem at $\tau = \tau'$, it suffices to check that the function $\bar{M}(\tau', \Lambda)$, minimum over $\Lambda \in \Omega$, is attained if and only if $\Lambda = \Lambda^*$.

Under a fixed Λ, introduce the notation

$$\Lambda_{t,\Delta} = (\lambda_1, \dots, \lambda_{k-t-1}, \lambda_{k-t} + \Delta, \dots, \lambda_k + \Delta)^T,$$
$$\psi_t(\Delta) = \bar{M}(\tau', \Lambda_{t,\Delta}), \quad t = 1, 2, \dots, k.$$

Note the evident equality

$$\frac{\partial}{\partial \Delta} \psi_t(\Delta) \Big|_{\Delta=0} = \sum_{j=k-t}^{k} \frac{\partial}{\partial \lambda_j} \bar{M}(\tau', \Lambda).$$

Suppose that for $\Lambda \in \Omega$

$$\frac{\partial}{\partial \Delta} \psi_t(\Delta) \Big|_{\Delta=0} \leq 0, \quad t = 1, 2, \dots, k, \tag{5.6}$$

and $\Delta = 0$ is not a point of the local minimum of $\psi_t(\Delta)$.

By virtue of the following Lemma, conclusion of the theorem for $\tau = \tau'$ follows.

Lemma 5.1. *Let the function $f(\Lambda) = f(\lambda_1, \dots, \Lambda_k)$ be defined and continuous on the set Ω and let the partial derivatives $\partial f(\Lambda) / \partial \lambda_i$ $(i = 1, \dots, k,\ \Lambda \in \Omega)$ exist. Denote $\varphi_t(\Delta) = f(\lambda_1, \dots, \lambda_{k-t-1}, \lambda_{k-t} + \Delta, \dots, \lambda_k + \Delta)$. For any $\Delta \in \Omega$, let $\frac{\partial}{\partial \Delta}\varphi_t(\Delta) \leq 0$, $t = 1, 2, \dots, k$ and $\Delta = 0$ is not a local minimum point of $\varphi(\Delta)$. Then $f(\Lambda') = \min_{\Lambda \in \Omega} f(\Lambda)$ if and only if $\Lambda' = \Lambda^*$, where Λ^* is defined in the statement of Theorem 5.1.*

Proof. Evidently

$$\sum_{j=k-1}^{k} \frac{\partial f(\Lambda)}{\partial \lambda_j} = \frac{\partial}{\partial \Delta} \varphi_t(\Delta) \Big|_{\Delta=0} \leq 0, \quad t = 1, \dots, k.$$

Using this inequality, we have

$$f(\Lambda) = f(\lambda_1, \dots, \lambda_{k-1}, \lambda_k) \geq f(\lambda_1, \dots, \lambda_{k-1}, \lambda_k + (\lambda_{k-1} - \lambda_k - \rho_{k-1})) \geq$$
$$\geq f(\lambda_1, \dots, \lambda_{k-2}, \lambda_{k-1} + (\lambda_{k-2} - \lambda_{k-1} - \rho_{k-2}),$$
$$\lambda_k + (\lambda_{k-1} - \lambda_k - \rho_{k-1}) + (\lambda_{k-2} - \lambda_{k-1} - \rho_{k-2})) \geq \dots$$
$$\dots \geq f(\lambda_1, \lambda_1 - \rho_1, \dots, \lambda_1 - \sum_1^{k-1} \rho_i) \geq f(\Lambda^*).$$

Here equality is attained if and only if $\Lambda = \Lambda^*$. The lemma is complete.

To prove (5.6), let us investigate a simpler determinant first: $J(\Theta) = \det \| e^{-\theta_1 x_i}, \dots, e^{\theta_m x_i} \|_{i=1}^{m}$ where $0 < x_1 < \dots < x_m$ are fixed, $\Theta = (\theta_1, \dots, \theta_m)^T$, $\theta_1 > \theta_2 > \dots > \theta_m$, m is an arbitrary natural number. Show that the inequalities

$$\sum_{j=m-t+1}^{m} J'_{\theta_j}(\Theta) < 0, \quad t = 1, 2, \dots, m \tag{5.7 }$$

are valid. Determine the function

$$g_m(z) = \begin{cases} J'_z(\Theta_z), & t = 1, \\ \sum_{j=m-t+1}^{m-1} J'_{\theta_j}(\Theta_z) + J'_z(\Theta_z), & 1 < t \leq m \end{cases} \tag{5.8}$$

where $\Theta_z = (\theta_1, \dots, \theta_{m-1}, z)^T$, $m > 1$, $\Theta_z = z$ for $m = 1$. The function $g_m(z)$ is defined by (5.8) for any fixed t. Evidently

$$\sum_{j=m-t+1}^{m} J'_{\theta_j}(\Theta) = g_m(\theta_m).$$

Hence, to prove the system of inequalities (5.7), it suffices to check the next assertion: inequality $g_m(\theta_m) < 0$ is valid for any fixed t, m, Θ such that $t < m$. This assertion proof will be carried out by induction on m. It is based on the investigation of oscillation properties of the function $g_m(z)$. For $m = 1$, we have $g_1(\theta_1) = -x_1 \exp(-\theta_1 x_1) < 0$.

Now assume that the statement being proved is valid for $m = m_0$ and show that it is valid for $m = m_0 + 1$. Expanding the determinants in (5.8) by a column depending on z, we have

$$g_{m_0+1}(z) = \sum_{p=1}^{m_0+1} b_p \exp(-x_p z),$$

where b_p $(p = 1, 2, \dots, m_0 + 1)$ is independent of z. Hence $g_{m_0+1}(z) \sim b_{m_0+1} e^{-x_{m_0+1} z}$ $(z \to -\infty)$, where

$$b_{m_0+1} = g_{m_0}(\theta_{m_0}) + J(\Theta)(-x_{m_0+1}) < 0, \ \Theta = (\theta_1, \dots, \theta_{m_0})^T.$$

In the remaining part of the proof, we will write g instead of g_{m_0+1} for a simplification of notation. Let $m = m_0 + 1$. For $t > 1$, $z = \theta_i$ $(i = m-t+1, \dots, m-1)$ all summands in (5.8), except two, are equal to zero since two columns in the corresponding determinants coincide. Similarly, if $z = \theta_i$ $(i = 1, 2, \dots, m - t)$, then all the summands in (5.8) except one are equal to zero. Therefore, for $t > 1$

$$\begin{aligned} g(\theta_i) = (-1)^{m-i-1}[&J'_{\theta_i}(\theta_1, \dots, \theta_i, z, \theta_{i+1}, \dots, \theta_{m-1}) + \\ &+ J'_z(\theta_1, \dots, \theta_i, z, \theta_{i+1}, \dots, \theta_{m-1})] \,|_{z=\theta_i} = 0 \end{aligned}$$
$$(i = m - t + 1, \dots, m - 1)$$

and for $t < m$

$$\begin{aligned} g(\theta_i) &= (-1)^{m-i-1} J'_z(\theta_1 < \dots, \theta_i, z, \theta_{i+1}, \dots, \theta_{m-1}) \,|_{z=\theta_i} \\ &= (-1)^{m-i-1} \det \| e^{-\theta_1 x_s}, \dots, e^{-\theta_i x_s}, -x_s e^{-\theta_i x_s}, e^{-\theta_{i+1} x_s}, \dots, e^{-\theta_{m-1} x_s} \|_{s=1}^{m} \\ &= (-1)^{m-i-1} c_i \ \ (i = 1, 2, \dots, m - t), \end{aligned}$$

where c_i are constants, $c_i < 0$ $(i = 1, \dots, m-t)$ by virtue of the Tchebyshev property of an exponential system. In these formulas, the degree $(-1)^{m-i-1}$ appears because the interchanging of a column depends on z.

Thus, the function $g(z)$ has the form

$$g(z) = \sum_{p=1}^{m_0+1} b_p \exp(-x_p z),$$

is negative for small z (since $b_{m_0+1} < 0$) and $g(\theta_i) = 0$ $(i = m-t+1, \dots, m-1)$. According to the Tchebyshev property of an exponential system, $g(z)$ cannot have more than m_0 roots. At $t = m = m_0 + 1$, $g(z)$ has roots $z = \theta_i$ $(i = 1, \dots, m-1)$, the general number of which is equal to m_0, hence $g(z) < 0$.

Consider the case of $t < m$. We have $(-1)^{t-1} g(\theta_{m-1}) < 0$. Moreover, the function $g(z)$ is negative for small z. Hence, in a set $W = (-\infty, \theta_{m-1}) \cup (\theta_{m-1}, \theta_{m-2}) \cup \dots \cup (\theta_{m-t+1}, \theta_{m-1})$, the function $g(z)$ has an even number of roots, since $g(\theta_i) = 0$ $(i = m-t+1, \dots, m-1)$. In addition, $g(z) = 0$ for some $z_s \in (\theta_s, \theta_{s+1})$, $s = 1, \dots, m-t-1$, as $g(z)$ has the opposite signs at these interval bounds, and is continuous. The number of zeros of function $g(z)$ enumerated above is equal to $m_0 - 1$. Hence, for $z \in W$, $g(z)$ cannot have more than one root, and since the number of roots on W must be even, it vanishes. Hence, $g(\theta_m)$ is negative for any $t (\leq m)$, $\theta_1 > \theta_2 > \dots > \theta_{m_0+1}$ which proves the statement.

Therefore, a set of inequalities (5.7) is complete. Let us go over to the proof of (5.6). Assume

$$\Lambda_{t,\delta,\Delta} = (\lambda_1, \lambda_1 + \delta, \dots, \lambda_{m-t} + \delta, \lambda_{m-t+1} + \Delta, \lambda_{m-t+1} + \Delta + \delta, \dots, \lambda_k + \Delta, \lambda_k + \Delta + \delta)^T,$$

$$g_t(\delta, \Delta) = J(-\Lambda_t, \delta, \Delta).$$

Elementary calculations show that

$$\begin{aligned} \psi_t(\Delta) &= \lim_{\delta \to 0} g_t(\delta, \Delta) / \delta^k, \\ \frac{\partial}{\partial \Delta} \psi_t(\Delta) \bigg|_{\Delta=0} &= \lim_{\delta \to 0} \frac{\partial}{\partial \Delta} g_t(\delta, \Delta) / \delta^k \bigg|_{\Delta=0}, \\ \frac{\partial}{\partial \Delta} g_t(\delta, \Delta) \bigg|_{\Delta=0} &= \sum_{j=1}^{2t} \frac{\partial}{\partial \theta_j} J(\tilde{\Theta}), \end{aligned} \tag{5.9}$$

where $\tilde{\Theta} = (\tilde{\theta}_1, \dots, \tilde{\theta}_{2k})^T$, $\tilde{\theta}_1 = \lambda_1$, $\tilde{\theta}_2 = \lambda_1 + \delta, \dots, \tilde{\theta}_{2k-1} = \lambda_k$, $\tilde{\theta}_{2k} = \lambda_k + \delta$.

According to (5.7), for sufficiently small $|\delta|$, $\delta < 0$

$$\sum_{j=1}^{2k} \frac{\partial}{\partial \Theta_j} J(\tilde{\theta}) < 0 \quad (t = 1, 2, \dots, k).$$

Therefore for $\Lambda \in \Omega,\ 0 \le x_1 < \ldots < x_{2k}$

$$\frac{\partial}{\partial \Delta}\psi_t(\Delta)\Big|_{\Delta=0} \le 0 \quad (t = 1, 2, \ldots, k),$$

$\Delta = 0$ is not a local minimum point of $\psi_t(\Delta)$, since by virtue of (5.9), $\psi_t(\Delta)$ is a nonincreasing function for $\Delta \in (\lambda_{k-t+1} - \lambda_{k-t}, 0)$.

Theorem 5.1 is thus proved for the case of $\tau = \tau' = \{x_1, \ldots, x_{2k}\}$.

For an arbitrary design $\tau = \{x_1, \ldots, x_N\}$ $(N > 2k)$, the reasoning shown above is to be applied to each summand in the right side of (5.3). The proof of Theorem 5.1 is now complete.

Using calculations similar to those allowing us to go over from (5.7) to (5.8), it can be shown that the statement of the theorem is valid for the information matrix determinant of the regression function

$$\sum_{i=1}^{k}\sum_{j=0}^{t_i} a_{ij}x^j \exp(-\lambda_j x),$$

where $\{a_{ij}, \lambda_j\}$ are the parameters under estimation.

Note that

$$\bar{M}(\tau, \Lambda) = \lim_{\delta \to 0} J(\widetilde{\Theta}) / \delta^k, \tag{5.10}$$

where $\widetilde{\Theta} = (\theta_1, \ldots, \theta_{2k})^T$, $\tilde{\theta}_1 = \lambda_1$, $\tilde{\theta}_2 = \lambda_2 + \delta, \ldots, \tilde{\theta}_{2k-1} = \lambda_k$, $\tilde{\theta}_{2k} = \lambda_k + \delta$. Using this relation, we have the inequality

$$\det M(\tau, A, \Lambda) \le a_1^2 \ldots a_k^2\ (2k)! \times$$
$$\times \sum_{1 \le \alpha_1 < \ldots < \alpha_{2k} \le N} \left[e^{-2\lambda_1 x_{\alpha_1} - 2\lambda_2 x_{\alpha_3} - \ldots - 2\lambda_k x_{\alpha_{2k-1}}} \right]^2 .$$

From this inequality, it follows that $\det M(\tau, A, \Lambda) \to 0$ for $x_i \to \infty$ for any fixed N, A, i, x_j, $(j = 1, \ldots, N,\ j \ne i)$, $\Lambda \in S = \{\Lambda \in R^k;\ \min \lambda_s > 0\}$.

Thus, we can consider that $\mathcal{X} = \mathcal{X}' = [0, b]$, where b is sufficiently large. The design τ^* for which

$$\det M(\tau^*, A, \Lambda^*) = \max_{\tau} \det M(\tau, A, \Lambda^*) \quad (a_i \ne 0,\ i = 1, 2, \ldots, k), \tag{5.11}$$

where maximum is taken over the set of all exact designs τ under fixed N, exists by virtue of the continuity of the function $\det M(\tau, A, \Lambda)$ and the compactness of the set $(\mathcal{X}')^N$. By Theorem 5.1, condition (5.11) is equivalent to the condition

$$\min_{\Lambda \in \Omega,\, A \in \Omega'} \det M(\tau^*, A, \Lambda) = \max_{\tau} \min_{\Lambda \in \Omega,\, A \in \Omega'} \det M(\tau, A, \Lambda),$$

where Ω' is any set from R^k not containing vectors with zero coordinates and Ω is defined by formula (5.4).

For approximate designs, going to the designs concentrated at a finite number of points and using the formula for the information matrix determinant, we have

$$\det M(\xi, A, \Lambda) = a_1^2 \dots a_k^2 \sum_{1\le\alpha_1<\dots<\alpha_{2k}\le n} p_{\alpha_1} \cdots p_{\alpha_{2k}} \times$$
$$\times \left[\det \left\| e^{-\lambda_1 x_{\alpha_i}}, -x_{\alpha_i} e^{-\lambda_1 x_{\alpha_i}}, \dots, e^{-\lambda_k x_{\alpha_i}}, -x_{\alpha_i} e^{-\lambda_k x_{\alpha_i}} \right\|_1^{2k}\right],$$

where $\xi = \{x_1, \dots, x_n;\ p_1, \dots, p_n\}$ ($\sum(p_i) = 1,\ p_i > 0$). Similarly, we have that the conditions

$$\det M(\xi^*, A, \Lambda) = \max \det M(\xi, A, \Lambda^*)$$

and

$$\min_{\Lambda\in\Omega,\, A\in\Omega'} \det M(\xi^*, A, \Lambda) = \max_{\xi} \min_{\Lambda\in\Omega,\, A\in\Omega'} \det M(\xi, A, \Lambda)$$

are equivalent at Ω, Ω' described above. Maximum in these expressions is taken over the set of all approximate designs. Thus, the problem of the determination of D-optimal designs at $\Lambda \in \Omega$, $A \in \Omega'$ is equivalent to that of the determination of locally D-optimal designs at $\Lambda = \Lambda^*$.

5.2. Saturated locally D-optimal designs.

We will now investigate designs for which the value

$$\det M(\tau, A, \Lambda)$$

attains its maximum under a fixed Λ (as is noted above, maximum location does not depend on A at $a_i \neq 0$ ($i = 1, \dots, k$)) in the class of all exact designs concentrated at $2k$ points ($\tau = \{x_1, \dots, x_{2k}\}$, $x_i \in \mathcal{X}$). These designs will be called saturated locally D-optimal designs. The problem of characterizing locally D-optimal designs (in the general sense, i.e., designs maximizing the information matrix determinant in the class of all approximate designs) is considerably more complicated and is beyond the scope of this work.

Using formula (5.3), we have that a saturated locally D-optimal design $\tau^* = (x_1^*, \dots, x_{2k}^*)$ is the design for which

$$\bar{M}(\tau^*, \Lambda) = \max \bar{M}(\tau, \Lambda) \tag{5.12}$$

is valid, where

$$\bar{M}(\tau, \Lambda) = (-1)^k \det \left\| e^{-\lambda_1 x_i}, -x_i e^{-\lambda_1 x_i}, \dots, e^{-\lambda_k x_i}, -x_i e^{\lambda_k x_i} \right\|_1^{2k},$$

and maximum is determined in the class of all exact designs with the number of points equal to $2k$. Our aim is to investigate $\tau^*(\Lambda)$ dependence. If Λ is such that $\lambda_i = \lambda_{i+1}$ at some $i = 1, \dots, k-1$, , then $\bar{M}(\tau, \Lambda) \equiv 0$.

For such Λ, equality (5.12) is valid for any τ^*. It turns out that (5.12) can be replaced with an equation having an unique solution for any Λ. If we select Λ such that $\lambda_1 < \ldots < \lambda_k$, then the new equation will be equivalent to (5.12).

Consider the function

$$\tilde{M}(\tau,\Lambda) = \bar{M}(\tau,\Lambda) \Big/ \prod_{i<j}(\lambda_i - \lambda_j)^4 . \tag{5.13}$$

Show that the definition of the function $\tilde{M}(\tau,\Lambda)$ can be extended with continuity remaining at the points Λ, such that $\lambda_i = \lambda_j$ for some i and j, $\quad i \neq j$. First, we have the following result.

Lemma 5.2. *The formula*

$$\begin{aligned}
\bar{M}(\tau,\Lambda) = {} & [2!\ldots(2k-1)!]^{-1} \exp\Big(-d\sum_1^{2k} x_i\Big) \times \\
& \times \prod_{1\le p<q\le k} (\lambda_p - \lambda_q)^4 \prod_{2k\ge t>l\ge 1} (x_t - x_l) \Big\{ 1 - \frac{1}{2k}\sum_1^k (\lambda_j - d) \times \\
& \times \sum_1^{2k} x_i + \frac{1}{2}\sum_1^k (\lambda_j - d)^2 \frac{\sum x_i^2(2k-1) - 2\sum_{i\neq j} x_i x_j}{2k(2k-1)(2k+1)} + \\
& + \frac{1}{2}\Big[\sum_1^k (\lambda_i - d)\Big]^2 \frac{\sum x_i^2(2k-1) + 4k\sum_{i\neq j} x_i x_j}{2k(2k-1)(2k+1)} + o(\alpha)\Big\},
\end{aligned} \tag{5.14}$$

where $\alpha = \max_i (\lambda_i - d)^2$, $\quad d$ is an arbitrary real number, $o(\alpha)/\alpha \to 0$ is valid.

Proof. Let $x_1, \ldots, x_m$ be fixed. Consider the determinant

$$J(\Theta) = \det \| \exp(\theta_i x_j) \|_{i,j=1}^m .$$

Using the exponent expansion in a Taylor series and the Binet–Cauchy formula, we have

$$\det \|\exp(-\theta_i x_j)\|_{i,j=1}^{m} =$$
$$= \det \left\| \sum_{s=0}^{m+1} (-1)^s (\theta_i x_j)^s \,/\, s! + o(\theta_i x_j)^{m+1} \right\|_{i,j=1}^{m} =$$
$$= \frac{(-1)^{[m/2]}}{2!\dots(m-1)!} \det \|x_j^{s-1}\|_{i,s=1}^{m} \det \|\theta_i^{s-1}\|_{i,s=1}^{m} +$$
$$+ \frac{(-1)^{[m/2]}}{2!\dots(m-2)!\, m!} \det \|x_j^{s-1}\|_{i\in 1:m,\, s\in 1:m+1,\, s\neq m} \times$$
$$\times \det \|\theta_i^{s-1}\|_{i\in 1:m,\, s\in 1:m+1,\, s\neq m} +$$
$$+ \frac{(-1)^{[m/2]}}{2!\dots(m-2)!\,(m+1)!} \det \|x_j^{s-1}\|_{i\in 1:m,\, s\in 1:m+2,\, s\neq m,m+1} \times$$
$$\times \det \|\theta_i^{s-1}\|_{i\in 1:m,\, s\in q:m+2,\, s\neq m,m+1} +$$
$$+ \frac{(-1)^{[m/2]}}{2!\dots(m-3)!\,(m-1)!\, m!} \det \|x_j^{s-1}\|_{j\in 1:m,\, s\in 1:m+1,\, s\neq n-1} \times$$
$$\times \det \|\theta_i^{s-1}\|_{i\in 1:m,\, s\in 1:m+1,\, s\neq m-1} +$$
$$+ o(\max_{r\in 1:m} \theta_r^{(m+1)\, m/2+2}).$$

We have the expressions for the determinants contained in this formula. The determinant

$$\det \|x_j^{s-1}\|_{j=1,s=1}^{m}$$

is a Vandermonde determinant and therefore

$$\det \|x_j^{s-1}\|_{j=1,s=1}^{m} = \prod_{m\geq p>q\geq 1} (x_p - x_q).$$

Similarly,

$$\det \|\theta_j^{s-1}\|_{j,\,s=1}^{m} = \prod_{m\geq p>q\geq 1} (\theta_p - \theta_q).$$

The determinant $\det \|x_j^{s-1}\|_{j\in 1:m,\, s\in 1:m+1,\, s\ m}$ is derived from $\det \|x_j^{s-1}\|_{j=1,s=1}^{m+1}$ by deleting a column, depending on x_{m+1}, and the last but one row. Using the determinant expansion in a row, we have

$$\det \|x_j^{s-1}\|_{j,s+1}^{m+1} = \sum_{j=1}^{m+1} (-1)^{j+m}\, x_j^{m-1} \times$$
$$\times \det \|x_i^{s-1}\|_{i,\, s\in 1:m+1,\, i\neq j,\, s\neq m} = \prod_{m+1\geq p>q\geq 1} (x_p - x_q).$$

Hence, equating polynomials termwise, we have

$$\det \left\| x_j^{s-1} \right\|_{j\in 1:m,\, s\in 1:m+1,\, s\neq m} = \sum_{i=1}^{m} x_i \prod_{m\geq p>q\geq 1} (x_p - x_q).$$

By a similar technique but using the expansion of the determinant with two rejected rows, we have

$$\det \left\| x_i^{s-1} \right\|_{i\in 1:m,\, s\in 1:m+2,\, s\neq m,m+1} =$$

$$= \left[\left(\sum_1^m x_i \right)^2 + \sum_{i\neq j} x_i x_j \right] \prod_{m\geq p>q\geq 1} (x_p - x_q),$$

$$\det \left\| x_i^{s-1} \right\|_{i\in 1:m,\, s\in 1:m+1,\, s\neq m-1} = \sum_{i\neq j} x_i x_j \prod_{m\geq p>q\geq 1} (x_p - x_q).$$

Certainly, the same formulas can be written for the determinants depending on Θ.

Using the equality

$$\sum_{i\neq j} x_i x_j = \frac{1}{2} \left(\sum_{i=1}^m x_i \right)^2 - \frac{1}{2} \sum_{i=1}^m x_i^2$$

and the obtained results, we get the formula

$$\begin{aligned} J(\Theta) = & \prod_{m\geq p>q\geq 1} (x_p - x_q) \prod_{m\geq t>l\geq 1} (\theta_t - \theta_l) \times \\ & \times [2! \ldots (m-1)!]^{-1} \left\{ 1 - \sum_1^m x_i \sum_1^m \theta_j \Big/ m \right\} + \\ & + \frac{1}{2} \sum_1^m \theta_i^2 \left\{ (m-1) \sum x_i^2 - 2 \sum_{i\neq j} x_i x_j \right\} \Big/ m(m-1)(m+1) + \\ & + \frac{1}{2} \left(\sum_1^m \theta_i \right)^2 \left[(m-1) \sum x_i^2 + 2m \sum_{i\neq j} x_i x_j \right] / m(m-1) \times \\ & \times (m+1) + o(\max_{r\in 1:m} \theta_r^2). \end{aligned} \tag{5.15}$$

Further,

$$\tilde{M}(\tau, \Lambda) = \lim_{\Delta\to 0} J(\tilde{\Lambda}) / \Lambda^k, \tag{5.16}$$

where $\tau = \{x_1, \ldots, x_{2k}\}$, $\tilde{\Lambda} = (\lambda_1, \lambda_1 + \Delta, \ldots, \lambda_k, \lambda_k + \Delta)^T$, $\lambda_1 > \lambda_2 > \ldots > \lambda_k$.

Assuming $m = 2k$, $\Theta = \tilde{\Lambda}$ in (5.15) and using (5.16), we have (5.14). The lemma is complete.

Lemma 5.3. I. *The equality*

$$\lim_{\Lambda \to \Lambda_d} \bar{M}(\tau, \Lambda) \Big/ \prod_{i<j} (\lambda_i - \lambda_j)^4 =$$

$$= \frac{1}{2! \ldots (2k-1)} \exp\left(-d \sum_{i=1}^{2k} x_i\right) \prod_{1 \le i < j \le 2k} (x_j - x_i),$$

takes place, where $\Lambda_d = (d, \ldots, d)^T \in R^k$, *d is an arbitrary real number.*

II. *The definition of the function* $\tilde{M}(\tau, \Lambda)$ *can be extended with continuity remaining at points* Λ *such that* $\lambda_i = \lambda_j$ *for some* $i \ne j$.

Proof. Statement I follows directly from formula (5.14).

II. Let Λ have the form $\Lambda = \Lambda' = (\lambda_1, \ldots, \lambda_{j-1}, \ d, \ldots, d, \ \lambda_{j+l}, \ldots, \lambda_k)^T$, where $\lambda_1 > \ldots > \lambda_{j-1} > d > \lambda_{j+l} > \ldots > \lambda_k$.

Represent the determinant $\bar{M}(\tau, \Lambda')$ in the form of the sum of the determinants of pairs of products, one of which depends only on $\lambda_1, \ldots, \lambda_{j-1}, \lambda_{j+l}, \ldots, \lambda_k$, while another depends on $\lambda_j, \ldots, \lambda_{j+l-1}$ $(\lambda_j = \ \ldots \ = \lambda_{j+l-1} = d)$. For the second determinants in these pairs, the same expansion as in Lemma 3.1 can be derived. Hence, it follows that the function $\tilde{M}(\tau, \Lambda)$, determined at the point $\Lambda = \Lambda'$ in the sense of passage to the limit, is continuous. Other cases are considered in the same way. The proof of Lemma 5.3 is complete.

Let $\tilde{M}(\tau, \Lambda)$ mean a function extended by the way indicated in Lemma 5.3. Let $S = \{-\Lambda \in R^k; \ \lambda_j > 0 \ (j = 1, 2, \ldots, k)\}$. Assume $\tau^*(\Lambda) = \{x_1^*(\Lambda), \ldots, x_{2k}^*(\Lambda)\}$.

Definition. Let a mapping $\tau^*(\Lambda) : S \to \mathcal{X}^{2k}$ $(\mathcal{X} = (0, +\infty))$ be called an optimal design function (ODF) for regression (5.2) if, for any $\Lambda \in S$,

$$\tilde{M}(\tau^*(\Lambda), \Lambda) = \max_{\tau} \tilde{M}(\tau, \Lambda),$$

where maximum is taken over all designs τ of the type $\tau = \{x_1, \ldots, x_{2k}\}$, $x_1 \le x_2 \le \ldots \le x_{2k}$, $x_i \in \mathcal{X}$ $(i = 1, 2, \ldots, 2k)$.

Later on we will show that ODF is defined in an unique way for any $\Lambda \in S$ and possesses some useful properties.

Since $\bar{M}(\tau, \Lambda)$ and $\tilde{M}(\tau, \Lambda)$ differ only in multiplier, independent of $x_1, \ldots, x_{2k}$, then at any fixed Λ such that $\lambda_1 > \ldots > \lambda_k$, $\tau^*(\Lambda)$ is a saturated locally D–optimum design.

Just as inequality (5.7) at $t = 1$, it can be proved that

$$\frac{\partial}{\partial x_1} \det \|\exp(-x_i \theta_j)\|_{i,j=1}^{m} < 0$$

for any fixed $0 < x_1 < \ldots < x_{2k}$, $\theta_1 > \theta_2 > \ldots > \theta_{2k}$.

From (5.16), we have that

$$\frac{\partial}{\partial x_1}\tilde{M}(\tau,\Lambda)\le 0 \quad (\tau=\{x_1,\dots,x_{2k})$$

and a point $x_1>0$ cannot be a local extremum point of $\tilde{M}(\tau,\Lambda)$ under every fixed $x_2,\dots,x_{2k};\ \ \lambda_1,\dots,\lambda_k \quad (x_1<\dots<x_{2k},\ \lambda_1>\dots>\lambda_k)$. Hence, the function $\tilde{M}(\tau,\Lambda)$ maximum with respect to x_1, is always attained at $x_1=0$. Thus, $x_1^*(\Lambda)\equiv 0$.

Now let us prove the following statement.

Lemma 5.4. *The value $\tau^*(\Lambda)$ at the point Λ_d (where $\Lambda_d=(d,\dots,d)^T\in R^k,\ d>0$) is defined uniquely and $\tau^*(\Lambda_d)=\Gamma_d=\{0,\gamma_2,\dots,\gamma_{2k}\},\ \ \gamma_i=2\gamma'_{i+1}\,/\,d$, where $\{\gamma'_i\}_{i=1}^{2k-1}$ are roots of $L^1_{2k-1}(x)$ (Laguerre polynomial with a parameter $\alpha=1$).*

Proof. We have proved that $x_1^*(\Lambda)\equiv 0$. Assume that $x_1=0$ and $\sum_2^{2k}x_i=D>0$. Then the maximum of the value

$$\tilde{M}(\tau,\Lambda_d)=\frac{1}{2!\dots(2k-1)!}\exp(-dD)\prod_{i=2}^{2k}x_i\prod_{1<q<p\le 2k}(x_p-x_q) \tag{5.17}$$

is attained if and only if $\{x_{v+1}\}_{v=1}^{k-1}$ is a collection of roots of the polynomial $L^1_{2k-1}(D^{-1}\,2k(2k-1)x)$ [47, ch. 10]. Moreover, the collection of these roots is the unique stationary point of the function $\tilde{M}(\tau,\Lambda_d)$ under the fixed d, since the logarithm of the right side of (5.17) is a convex function τ. Further, with the help of direct differentiation, we can see that the maximum over D is attained at $D=(2k-1)\,k/d$. The proof of the lemma is complete.

If a design $\tau=\{x_1,\dots,x_{2k}\}$ contains at least two coinciding points, then $\tilde{M}(\tau,\Lambda)\equiv 0$. Hence, $0<x_2^*(\Lambda)<\dots<x_{2k}^*(\Lambda)$ and $d\tilde{M}(\tau^*(\Lambda),\Lambda)\,/\,dx_i=0 \quad (i=1,2,\dots,2k)$. Thus, the optimal design function $\tau^*(\Lambda)$ satisfies the relations

$$\begin{aligned}&x_1^*(\Lambda)=0,\\&\frac{\partial}{\partial x_i}\tilde{M}(\tau^*(\Lambda),\Lambda)=0,\quad i=1,2,\dots,2k,\\&0<x_2^*(\Lambda)<\dots<x_{2k}^*(\Lambda).\end{aligned} \tag{5.18}$$

Relation system (5.18) implicitly defines the vector-function $\tau^*(\Lambda)$. Let us provisionally use this notation for every solution of (5.18). Later it will be clear that for $\Lambda\in S$ such a solution exists, is unique, and coincides with the ODF.

Lemma 5.5. *A matrix*

$$\left\|\frac{\partial^2}{\partial x_i\,\partial x_j}\tilde{M}(\tau^*(\Lambda),\Lambda)\right\|_{i,j=2}^{2k},$$

where $\tau^*(\Lambda)$ satisfies (5.18) and $\Lambda \in S$, is nonnegatively defined.

Proof. Throughout the proof, τ has the form $\tau = \{x_1, \ldots, x_{2k}\}$, $0 \le x_1 < \ldots < x_{2k}$, $\Lambda \in S$. Let us start with the statement that if τ and Λ are such that $\frac{\partial}{\partial x_i}\tilde{M}(\tau, \Lambda) = 0$ for some fixed i ($i = 2, \ldots, 2k$), then

$$\frac{\partial^2}{\partial x_i^2} \tilde{M}(\tau, \Lambda) < 0.$$

Set $g(z) = \frac{\partial}{\partial z}\tilde{M}(\tau_z, \Lambda)$ where $\tau_z = \{x_1, \ldots, x_{i-1}, z, x_{i+1}, \ldots, x_{2k}\}$. Direct checking shows that $\tilde{g}(z) = \tilde{M}(\tau_z, \Lambda)$ vanishes at points $x_1, x_2, \ldots, x_{i-1}, x_{i+1}, \ldots, x_{2k}$. Hence, $g(z)$ vanishes on intervals $(x_1, x_2), \ldots, (x_{i-2}, x_{i-1}), (x_{i+1}, x_{i+2}), \ldots, (x_{2k-1}, x_{2k}), (x_{2k}, +\infty)$ and $g(x_i) = 0$. In addition, $g(z)$ has the form

$$\sum_{i=1}^{k} b_i \exp(-\lambda_i z) + \sum_{i=1}^{k} b_{i+k}\, z \exp(-\lambda_i z)$$

and by virtue of the Tchebyshev property of an exponential system, cannot have more than $2k - 1$ zeros (with regard to multiplicity). A simple count of zeros show that a point $z = x_i$ cannot be zero of 2 or more multiplicity for the function $g(z)$. Since $\tilde{g}(z) = \tilde{M}(\tau_z, \Lambda)$ vanishes for $z = x_{i-1}$, x_{i+1} and is strictly more than zero (by virtue of the Tchebyshev property of an exponential system) for $z \in (x_{i-1}, x_{i+1})$, $g(z) = \tilde{g}'(z)$ has a unique zero for $z \in (x_{i-1}, x_{i+1})$ at a point $z = x_i$, then the point $z = x_i$ is the point of function $\tilde{g}(z)$ local maximum and

$$\frac{\partial^2}{\partial x_i^2} \tilde{M}(\tau, \Lambda) = \frac{\partial}{\partial z} g(z)\Big|_{z=x_i} < 0,$$

Q.E.D.

Now prove the statement: if τ and Λ are such that $\frac{\partial}{\partial x_i}\tilde{M}(\tau, \Lambda) = 0$ for a fixed i ($i = 2, 3, \ldots, 2k$), then $\frac{\partial^2}{\partial x_i \partial x_j}\tilde{M}(\tau, \Lambda) > 0$ at any $j \ne i$.

Set $g(z) = \frac{\partial}{\partial x_i}\tilde{M}(\tau_z, \Lambda)$, where $\tau_z = \{x_1, \ldots, x_{j-1}, z, x_{j+1}, \ldots, x_{2k}\}$. Note that $g(x_l) = 0$ at $l = 1, \ldots, 2k$, $l \ne i$ since two rows (at $l \ne j, i$) coincide in the corresponding determinants and according to the statement $g(x_i) = 0$.

Moreover, $g(z) \to 0$ at $z \to \infty$, since all the units of some column of the corresponding determinant tend to zero, and the remaining units are constant. Hence, the function $g'(z)$ vanishes at the intervals $(x_1, x_2), \ldots, (x_{i-2}, x_{i-1}), (x_{i+1}, x_{i+2}), \ldots, (x_{2k-1}, x_{2k}), (x_{2k}, \infty)$. Counting zeros and using the Tchebyshev property of an exponential system, we have $g'(x_j) = 0$. Further investigating the sign changing of the function $g(z)$, we have $\operatorname{sign} g'(z)|_{z=x_j} = \operatorname{sign} g(x_j)(-1)^{j-i+1}h$, where $h = 1$ if $j < i$ and $h = -1$ if $j > i$.

Besides

$$h\, g(x_i)(-1)^{j-i+1} = \lim_{\Delta \to 0} \tilde{M}(\tau_{\Delta}, \Lambda) / \Delta,$$

where $\tau_{\Delta} = \{x_1, \dots, x_i, x_i + \Delta, x_{i+1}, \dots, x_{j-1}, x_{j+1}, \dots, x_{2k}\}$, $h = -1$ if $j > i$ and $\tau_{\Delta} = \{x_1, \dots, x_{j-1}, x_{j+1}, \dots, x_i, x_i + \Delta, x_{i+1}, \dots, x_{2k}\}$, $h = 1$ if $j < i$. By virtue of the Tchebyshev property of an exponential system, $\tilde{M}(\tau_{\Delta}, \Lambda) > 0$ at $\Delta \in (0, x_{i+1} - x_i)$, hence $g'(x_j) \geq 0$. Since $g'(x_j) \neq 0$ as was shown above, we have

$$\frac{\partial^2}{\partial x_i \partial x_j} \tilde{M}(\tau, \Lambda) = g'(x_j) > 0$$

Q.E.D.

Similarly, we can show that from $\frac{\partial}{\partial x_i} \tilde{M}(\tau, \Lambda) = 0$ under a fixed $i = 2, \dots, 2k$, it follows that $\frac{\partial^2}{\partial x_i \partial \lambda} \tilde{M}(\tau, \Lambda) > 0 \quad (j = 1, 2, \dots, k)$. This assertion is necessary for the proof of Lemma 5.6.

Let us prove the following identity

$$\sum_{i=1}^{2k} \frac{\partial}{\partial x_i} \tilde{M}(\tau, \Lambda) \equiv -2 \left(\sum_{j=1}^{k} \lambda_i \right) \tilde{M}(\tau, \Lambda). \tag{5.19}$$

In fact,

$$\det \| e^{-\lambda_k (x_i + \Delta)}, -(x_i + \Delta) e^{-\lambda_k (x_i + \Delta)}, \dots\dots, e^{-\lambda_k (x_i + \Delta)}, -(x_i + \Delta) e^{-\lambda_k (x_i + \Delta)} \|_1^{2k} =$$
$$= e^{-2 \sum_1^k \lambda_i \Delta} \det \| e^{-\lambda_1 x_i}, -x_i e^{-\lambda_1 x_i}, \dots, e^{-\lambda_k x_i}, -x_i e^{-\lambda_k x_i} \|_1^{2k}.$$

Differentiating this equality with respect to Δ and taking into account that

$$\frac{\partial}{\partial \Delta} \tilde{M}(\tau_{\Delta}, \Lambda) = \sum_{i=1}^{2k} \frac{\partial}{\partial x_i} M(\tau_{\Delta}, \Lambda),$$

where $\tau_{\Delta} = \{x_1 + \Delta, \dots, x_{2k} + \Delta\}$, we have (5.19).

From (5.19), we conclude that at any $\Lambda \in S$,

$$\frac{\partial}{\partial x_j} \sum_{i=1}^{2k} \frac{\partial}{\partial x_i} \tilde{M}(\tau^*(\Lambda), \Lambda) =$$
$$= -2 \left(\sum_{l=1}^{k} \lambda_l \right) \frac{\partial}{\partial x_j} \tilde{M}(\tau^*(\Lambda), \Lambda) = 0 \quad (j = 1, \dots, 2k)$$

where $\tau^*(\Lambda)$ is any solution of (5.18).

Thus, at a point $(\tau^*(\Lambda), \Lambda)$ we have (for brevity, we will omit the function $\tilde{M}$ argument)

$$\sum_{i \in 1:2h,\, i \neq j} \left| \frac{\partial^2}{\partial x_i \partial x_j} \tilde{M} \right| = \sum_{i \in 1:2h,\, i \neq j} \frac{\partial^2}{\partial x_i \partial x_j} \tilde{M} = -\frac{\partial^2}{\partial x_j^2} \tilde{M}$$

(we used the fact that $\frac{\partial^2}{\partial x_i \partial x_j}\tilde{M} > 0$ as was proved above).

Thus

$$\sum_{i \in 2:2k,\, i \neq j} \left| \frac{\partial^2}{\partial x_i \partial x_j} \tilde{M} \right| < \left| \frac{\partial^2}{\partial x_i^2} \tilde{M} \right| \leq \left| \frac{\partial^2}{\partial x_i^2} \tilde{M} - \mu \right| \tag{5.20}$$

for any $\mu \geq 0$, since $\partial^2 \tilde{M} \,/\, \partial x_i^2 < 0$.

Hence, according to the Hadamard's criterion

$$(-1)^{2k-1} \det[P - \mu I] > 0, \tag{5.21}$$

where I is an identity matrix, $\mu \geq 0$,

$$P = \left\| \frac{\partial^2}{\partial x_i \partial x_j} \tilde{M}(\tau^*(\Lambda), \Lambda) \right\|_{i,j=2}^{2k}.$$

Matrix P is a symmetric matrix and, hence, all its eigenvalues are real and, by virtue of (5.21), are negative. Therefore, the matrix P is negatively defined. The proof of the lemma is complete.

Denote $L = \{\tau;\, \tau = \{0, x_2, \ldots, x_{2k}\},\ 0 < x_2 < \ldots < x_{2k}\}$. Let $\tau' \in L$ and $\Lambda \in S$ be such that

$$\frac{\partial^2}{\partial x_i} \tilde{M}(\tau', \Lambda') = 0 \quad (i = 2, 3, \ldots, 2k).$$

Then, by virtue of Lemma 5.5,

$$\det \left\| \frac{\partial^2}{\partial x_i \partial x_j} \tilde{M}(\tau', \Lambda') \right\|_{i,j=2}^{2k} \neq 0.$$

By the implicit mapping theorem, in some vicinity (W) of the point Λ', we can determine a continuous vector function $\tilde{\tau}^*(\Lambda) = \{0, \tilde{x}_2^*(\Lambda), \ldots, \tilde{x}_{2k}^*(\Lambda)\}$ such that

$$\frac{\partial}{\partial x_i} \tilde{M}(\tau^*(\Lambda), \Lambda) = 0 \quad (i = 2, \ldots, 2k)$$

at $\Lambda \in W$ and $\hat{\tau}^*(\Lambda') = \tau'$. Let H be a vicinity of point Λ' such that $0 < \hat{x}_2^*(\Lambda) < \ldots < \tilde{x}_{2k}^*(\Lambda)$ at $\Lambda \in H$. Hence, for any fixed point $L' \in S$, there exists a vicinity $(W \cap H \cap s)$ where a vector function $\tilde{\tau}^*(\Lambda) = \{0, \tilde{x}_2^*(\Lambda), \ldots, \tilde{x}_{2k}^*(\Lambda)\}$ with the properties

$$\begin{aligned}
&(a)\ \ \partial \tilde{M}(\tilde{\tau}^*(\Lambda), \Lambda) \,/\, \partial x_i = 0 \ \ (i = 2, \ldots, 2k),\\
&(b)\ \ \tilde{\tau}^*(\Lambda') = \tau',\\
&(c)\ \ \tilde{\tau}^*(\Lambda) \text{ continuous},\\
&(d)\ \ 0 < \tilde{x}_2^*(\Lambda) < \ldots < \tilde{x}_{2k}^*(\Lambda),
\end{aligned}$$

is defined, where acccording to the implicit mapping theorem, a vector function with properties (a)–(d) is uniquely defined.

Consider a union $(\widetilde{W})$ of all vicinities of a point $\Lambda' = (\lambda_1', \ldots, \lambda_k')^T$, for which a vector function $\tau^*(\Lambda)$ with properties (a)–(d) can be defined. Evidently, (a)–(d) are valid at $\Lambda \in \widetilde{W}$.

Lemma 5.6. *Functions* $x_i^*(\Lambda)$ $(i = 2,\dots,2k)$ *strictly decrease with respect to each* λ_j $(j = 1,2,\dots,k)$ *at* $\lambda \in \widetilde{W}$.

Proof. By virtue of the superimposed conditions, the function $\tau^*(\Lambda)$ at $\Lambda \in \widetilde{W}$ is a solution of the system (5.18). Hence, the matrix

$$P = \left\| \frac{\partial^2}{\partial x_i \partial x_j} \tilde{M}(\tilde{\tau}^*(\Lambda), \Lambda) \right\|_{i,j=2}^{2k}$$

at $\Lambda \in \widetilde{W}$ has the form $P = \|p_{ij}\|_{i,j=2}^{2k}$, $p_{ii} < 0$, $p_{ij} > 0$ $(i \neq j)$ and is negatively defined (see the proof of Lemma 3.4). It is known [96] that in this case, all units of the matrix P^{-1} are negative. Since

$$\frac{\partial^2}{\partial \lambda_j \partial x_i} \tilde{M}(\tilde{\tau}^*(\Lambda), \Lambda) > 0 \quad (j = 1,2,\dots,k;\ i = 2,\dots,2k).$$

(see the proof of Lemma 5.5), then from the well-known formula for implicit mapping derivatives, it follows that $\partial \tilde{x}_i^*(\Lambda) / \partial \lambda_j < 0$ $(\Lambda \in \widetilde{W},\ i = 2,\dots,2k;\ j = 1,\dots,k)$. The proof of the lemma is now complete.

Denote a set $V = \{\Lambda \in R^k;\ \Lambda = (\lambda_1,\dots,\lambda_k)^T,\ \lambda_i \geq \lambda_i'\ (i = 1,2,\dots,k)\}$, $((\lambda_1',\dots,\lambda_k')^T = \Lambda')$ by V. Let us prove that $V \subset \widetilde{W}$, which is equivalent to $\lambda(\alpha) = (\lambda_1' + \alpha l_1,\dots,\lambda_k' + \alpha l_k)^T \in \widetilde{W}$ at any $\alpha \geq 0$ under the fixed $l_j \geq 0$ $(j = 1,2,\dots,k)$. Suppose, to the contrary, that this is not so. Then there exist α' and $\lambda_1,\dots,\lambda_k$ such that $\Lambda(\alpha') \notin \widetilde{W}$, $\Lambda(\alpha) \in \widetilde{W}$ $(0 \leq \alpha < \alpha')$. By virtue of property (d), functions $\tilde{x}_i^*(\Lambda(\alpha))$ are bounded from below by zero at $\alpha \in (0,\alpha')$ and, according to Lemma 5.6, decrease by α. Hence, there exists a finite limit $\lim_{\alpha \to \alpha'} \tilde{\tau}^*(\Lambda(\alpha))$. Denote this limit by $\tilde{\tau}(\Lambda(\alpha'))$. By virtue of the continuity of functions $\partial \tilde{M}(\tau,\Lambda) / \partial x_i$ with the respect to the collection of variables τ and Λ, we have $\partial \tilde{M}(\tilde{\tau}^*(\Lambda(\alpha'), \Lambda(\alpha')) / \partial x_i = 0$. Moreover, $0 \leq \tilde{x}_2^*(\Lambda(\alpha')) \leq \dots \leq \tilde{x}_{2k}^*(\Lambda(\alpha))$. Let us prove that $0 < \tilde{x}_2^*(\Lambda(\alpha')) < \dots < \tilde{x}_{2k}^*(\Lambda(\alpha))$. On the contrary, if $\tilde{x}_s^*(\Lambda(\alpha')) = \dots = \tilde{x}_{s+t}^*(\Lambda(\alpha'))$ at some s and t, then for any $\varepsilon > 0$ there exists $\delta > 0$ such that at $\Lambda = (\lambda_1,\dots,\lambda_k)^T \in \widetilde{W}$, $\max_i |\lambda_i - (\lambda_i' + \alpha' l_i)| < \delta$ and $\tau = \tau^*(\Lambda)$ we have $0 < \hat{x}_{j+1}^*(\Lambda) - \tilde{x}_j^*(\Lambda) < \varepsilon$ $(j = s, s+1,\dots,s+t-1)$ and $\partial \tilde{M}(\tau,\Lambda)/\partial x_i = 0$ $(i = 2,\dots,2k)$.

Consider the function

$$\tilde{\tilde{M}}(\tau,\Lambda) = \tilde{M}(\tau,\Lambda) \Big/ \prod_{1 \leq i < j \leq 2k} (x_j - x_i)$$

As in Lemma 5.3, we can determine this function with preserving continuity and continuous differentiability with respect to $x_1,\dots,x_{2k}$. Note that $\tilde{M}(\tau,\Lambda) \neq 0$ at any τ, Λ.

On the other hand,

$$\frac{\partial}{\partial x_l}\tilde{\tilde{M}}(\tau,\Lambda) = \Bigg\{ - \sum_{p\in 1:2k,\ p\neq l} \frac{1}{x_l - x_p}\tilde{\tilde{M}}(\tau,\Lambda) +$$

$$+\frac{\partial}{\partial x_l}\tilde{M}(\tau,\Lambda) \Big/ \prod_{i<j}(x_j - x_i)\Bigg\} = - \sum_{p\in 1:2k,\ p\neq l} \frac{1}{x_l - x_p}\tilde{\tilde{M}}(\tau,\Lambda),$$

since $d\tilde{M}(\tau,\Lambda)\,/\,dx_l = 0 \quad (l = 2,\dots,2k)$. Hence, there exists a sequence $\left\{\Lambda^{(s)}\right\}_{s=1}^{\infty}$, $\Lambda^{(s)} \to \Lambda(\alpha')$ at $s \to \infty$ such that

$$\frac{\partial}{\partial x_l}\tilde{\tilde{M}}\left(\tau(\Lambda^{(s)}),\ \Lambda^{(s)}\right) \to \infty \text{ at } s \to \infty.$$

This contradicts the continuous differentiability of the function $\tilde{\tilde{M}}(\tau,\Lambda)$. Hence, $0 < \tilde{x}_2^*(\Lambda(\alpha')) < \dots < \tilde{x}_{2k}^*(\Lambda(\alpha'))$.

Thus, conditions (a)–(d) are valid at $\Lambda = \Lambda(\alpha')$ for a continuous function $\tilde{\tau}^*(\Lambda)$ extended at a point $\Lambda = \Lambda(\alpha')$ and, hence, $\Lambda(\alpha') \in \widetilde{W}$. The obtained contradiction (it was assumed that $\Lambda(\alpha') \notin \widetilde{W}$) shows that $V \subset \widetilde{W}$.

Taken as a Λ' point of the form $\Lambda_d = (d,\dots,d)^T \in R^k,\ \ d = \varepsilon > 0$, we have that a function $\tilde{\tau}^*(\Lambda)$ with properties (a)–(d) can be determined on the set $S_\varepsilon = \left\{\Lambda \in R^k;\ \Lambda = (\lambda,\dots,\lambda_k)^T,\ \lambda_i \geq \varepsilon\ (i = 1,2,\dots,k)\right\}$ at any $\varepsilon > 0$.

The following lemma is needed.

Lemma 5.7. *The solution of system (5.18) is unique at any* $\Lambda \in S$.

Proof. The proof will be carried out by contradiction. Let, at some $\Lambda' \in S$, there exist $\tau_2 \neq \tau_1,\ \ \tau_1, \tau_2 \in L = \{\tau;\ \tau = \{0, x_1,\dots,x_{2k}\},\ 0 < x_2 < \dots < x_{2k}\}$ such that

$$\frac{\partial}{\partial x_i}\tilde{M}(\tau,\Lambda') = 0 \quad (i = 2,\dots,2k)$$

at $\tau = \tau_1, \tau_2$. As was shown above, continuous vector functions $\tau^{(1)}(\Lambda)$ and $\tau^{(2)}(\Lambda)$, such that $\tau^{(1)}(\Lambda') = \tau_1,\ \ \tau^{(2)}(\Lambda') = \tau_2,\ \ d\tilde{M}\left(\tau^{(i)}(\Lambda,),\Lambda\right)/\,dx_j = 0 \ \ (i = 1,2,\ j = 2,\dots,2k)$ and $\tau^{(i)}(\Lambda) \in L$ at any fixed $\Lambda \in V \ \ (i = 1,2)$, can be defined at $V = \{\Lambda;\ \Lambda = (\lambda_1,\dots,\lambda_k)^T,\ \lambda_i \geq \lambda_i'\ (i = 1,2,\dots,k)\}$. By virtue of Lemma 5.4, $\tau^{(1)}(\Lambda_d) = \tau^{(2)}(\Lambda_d)$ at $d = \max_{i\in 1:k}\lambda_i$. Therefore, we can find a point (denote it by $\tilde{\Lambda}$) in the vicinity of which two coinciding vector functions $\tau^{(1)}(\Lambda)$ and $\tau^{(2)}(\Lambda)$ are defined, such that $\tau^{(1)}(\tilde{\Lambda}) = \tau^{(2)}(\tilde{\Lambda})$. It contradicts the assertion of the uniqueness in the implicit mapping theorem. The obtained contradiction demonstrates that the function $\tau^*(\Lambda)$, i.e., the solution of (5.18), is determined uniquely at any $\Lambda \in S$. The proof of the lemma is complete.

So, the solution of (5.18) is unique and therefore it coincides with the ODF. As was shown above, function $\tau^*(\Lambda)$ is continuous on S. From the known explicit formulas for implicit vector function derivatives, we conclude that $x_i^*(\Lambda)$ $(i = 2,\dots,2k)$ are analytical functions for $\Lambda \in S$ (derivatives $(x_i^*(\Lambda))'_{\lambda_i}$ are analytical as fractions in which the numerator and denominator are analytical functions, and the denominator is equal to

$$\det \left\| \frac{\partial^2}{\partial x_i \partial x_j} \tilde{M}(\tau^*(\Lambda), \Lambda) \right\|_{i,j=2}^{2k},$$

by Lemma 5.5 does not vanish). Thus the following result is valid

Lemma 5.8. *The ODF is uniquely defined, and the functions $x_i^*(\Lambda)$ $(i = 2,\dots,2k)$ are analytic for $\lambda \in S$.*

The results obtained can be expressed in the form of the following theorem.

Theorem 5.2. *An optimal design function*

$$\tau^*(\Lambda) = \{x_1^*(\Lambda),\dots,x_{2k}^*(\Lambda)\} \;:\; S \to [0,+\infty]^{2k}$$

exists and is uniquely defined. Moreover, $x_1^*(\Lambda) \equiv 0$, $x_i^*(\Lambda)$ $(i = 2,\dots,2k)$ are analytical functions strictly decreasing in respect to every λ_j $(j = 1,\dots,k)$. Here $\tau^*(\Lambda_d) = \Gamma_d$, where $\Lambda_d = (d,\dots,d)^T \in (0,+\infty)^k$ and Γ_d is defined in Lemma 5.4.

Design $\{x_1^*(\Lambda),\dots,x_{2k}^*(\Lambda)\}$ is a saturated locally D-optimal design for the regression (5.2) at any fixed $\Lambda \in S$ such that $\lambda_1 < \lambda_2 < \dots < \lambda_k$.

Remark 1. Make a note touching upon saturated D-minimax designs. Assume $\Omega = \{\Lambda;\ \lambda_i - \lambda_{i+1} \geq \rho_i > 0\ (i = 1,2,\dots,k-1)\ \lambda_1 \geq d\}$. Let $\xi(\{\rho_i\})$ be a saturated D–minimax design with respect to a set Ω. Then, according to Theorem 5.1, $\xi = \{x_1^*(\Lambda),\dots,x_{2k}^*(\Lambda)\}$, where $\Lambda = \left((d, d-p,\dots,d-\sum_1^{k-1}\rho_i)\right)^T$.

Suppose that $\rho_i \to 0$ at $i = 1,2,\dots,k-1$. By continuity of the optimal design function, we have that

$$\xi(\{\rho_i\}) \to \{x_1^*(\Lambda_d),\dots,x_{2k}^*(\Lambda_d\})$$

and the form of the limit design is defined by Lemma 5.4.

Remark 2. Let a set $\mathcal{X}$, from which experimental points are selected, coincide with $[c,+\infty)$ where $c > 0$. Note that

$$\tilde{M}(\tau,\Lambda) = \exp\left(-2\sum_1^k \lambda_i c\right) \tilde{M}(\tau_c,\Lambda),$$

where $\tau = \{x_1,\dots,x_{2k}\}$, $\tau_c = \{x_1 - c,\dots,x_{2k} - c\}$. Therefore, a saturated locally D–optimal experiment design for $\mathcal{X} = [c,+\infty)$ under a fixed Λ is received by the translation of all design points in a constant c from a saturated locally D–optimal experiment design for $\mathcal{X} = [0,+\infty)$ under the same fixed Λ.

The following algorithm is recommended for the search of optimal points $x_2^*(\Lambda),\dots,$ $x_{2k}^*(\Lambda)$ under fixed Λ. As practical calculations show, this is efficient.

Algorithm. Let $x_1 = 0 < x_2 < \ldots < x_{2k} < x_{2k+1} = b$ be given. Define

$$x_{2k+1}^{(1)} = x_{2k+1}, \quad x_j^{(1)} = z_j^* \quad (j = 2k, \ldots, 2),$$

where z_j^* is such that

$$\bar{M}\left(\tau_{z_j^*}, \Lambda\right) = \max_{z \in \left[x_{j-1}, x_{j+1}^{(1)}\right]} \bar{M}(\tau_z, \Lambda), \tag{5.22}$$

where $\tau_z = \left\{x_1, \ldots, x_{j-1}, z, x_{j+1}^{(1)}, \ldots, x_{2k}^{(1)}\right\}$. Similarly, $x_j^{(2)}, \ldots, x_j^{(s)}, \ldots$ $(j = 2, \ldots, 2k+1)$ are determined. It is simple to prove the existence of the designs subsequence $\left\{0, x_2^{(s_i)}, \ldots, x_{2k}^{(s_i)}\right\}$, $i = 1, 2, \ldots$, converging to the stationary point of the function $\bar{M}(\tau, \Lambda)$ under fixed $x_1 = 0$ and fixed Λ (if b is rather large). Since here the stationary point is unique and it is a minimum point of the function $\bar{M}(\tau, \Lambda)$, then we have $x_j^{(s_i)} \to x_j^*(\Lambda)$ $(i \to \infty,\ j = 2, \ldots, 2k)$.

Using the technique for counting zeros and taking into account the Tchebyshev property of an exponential system, we have that the solution of (5.22) is unique. The numerical implementation of this method is rather simple.

By virtue of inequality (5.20), the descent direction at each step of the algorithm is close to the direction of the antigradient that provides the high rate of convergence.

Theorem 5.3. *A locally D-optimal design with a finite number of points for regression (5.1) is concentrated in two points for $k = 1$, in four points for $k = 2$, and in no more than $[3k(k+1)/4]$ points for $k > 2$ (where $\lfloor a \rfloor$ is the integral part of a).*

The proof of Theorem 5.3 is similar to that of the theorem about the number of points for a D-optimal design in the case of polynomial regression in paper [47], but it is necessary to refer to the Tchebyshev property of an exponential system instead of reference to the Tchebyshev property of a polynomial.

For $k = 1$, direct calculations show that an optimal design has the form $\{0, 1/\lambda_1\}$. For $k = 2$, an optimal design is concentrated at $2k = 4$ points and it can be found by means of the algorithm suggested above.

Offer the table of locally optimal designs (Table I) calculated according to the program made up on the basis of the described algorithm for the case $k = 2$. It is easy to see that $x_i^*(\Lambda') = x_i^*(\Lambda)/h$ if $\Lambda' = (\lambda_1', \ldots, \lambda_k')^T$, $\lambda_j' = h\lambda_j$ $(j = 1, 2, \ldots, k)$. Therefore take $(\lambda_1 + \lambda_2)/2 = 1$.

Table I shows the τ^* dependence on $\Delta = (\lambda_1 - \lambda_2)/2$. The values of $\bar{M}(\tau^*(\Lambda), \Lambda)$ and $\tilde{M}(\tau^*(\Lambda), \Lambda)$ are also given.

For properties characteristics of the design corresponding to $\Delta = 0$
(denote it by τ^*; in accordance with Lemma 5.4 this design can be found analytically) indicate values of the $\bar{M}$ determinant for $\Lambda = \Lambda^{(1)} = \left(\lambda_1^{(1)}, \lambda_2^{(1)}\right)^T$, $\frac{1}{2}\left(\lambda_1^{(1)} - \lambda_2^{(1)}\right) =$

TAB. 1. Table of locally optimal designs

Δ	0	0.1	0.2	0.3	0.4
X_1	0	0	0	0	0
X_2	0.47	0.47	0.47	0.48	0.48
X_3	1.66	1.66	1.66	1.69	1.70
X_4	3.88	3.91	3.97	4.08	4.23
$\bar{M}$	0	$0.99 \cdot 10^{-5}$	$0.15 \cdot 10^{-3}$	$0.81 \cdot 10^{-3}$	$0.28 \cdot 10^{-2}$
$\tilde{M}$	0.09	0.09	0.10	0.10	0.11

Δ	0.5	0.6	0.7	0.8	0.9	0.95
X_1	0	0	0	0	0	0
X_2	0.48	0.48	0.49	0.49	0.49	0.50
X_3	1.74	1.80	1.91	2.06	2.36	2.70
X_4	4.47	4.98	5.77	7.43	12.57	22.39
$\bar{M}$	$0.78 \cdot 10^{-2}$	$0.19 \cdot 10^{-1}$	$0.47 \cdot 10^{-1}$	0.12	0.39	0.99
$\tilde{M}$	0.12	0.15	0.19	0.29	0.59	1.10

$\Delta^{(1)} = 0.5$ and at $\Lambda = \Lambda^{(2)} = \left(\lambda_1^{(2)}, \lambda_2^{(2)}\right)^T$, $\frac{1}{2}\left(\lambda_1^{(2)} - \lambda_2^{(2)}\right) = \Delta^{(2)} = 0.6$ (compare with the corresponding values from the table):

$$\bar{M}\left(\tau^*, \Lambda^{(1)}\right) = 0.75 \cdot 10^{-2}, \quad \bar{M}\left(\tau^*, \Lambda^{(2)}\right) = 0.17 \cdot 10^{-1}.$$

It is useful to compare the quality of design τ^* with that of designs with uniformly distant points (such designs are very often used in practice). Let
$\tau_1 = \{0, 4/3, 8/3, 4.0\}$, $\tau_2 = \{0, 2.5, 5.0, 7.5\}$, $\lambda = (\lambda_1, \lambda_2)^T$; $(\lambda_1 + \lambda_2)/2 = 1$, $(\lambda_1 - \lambda_2)/2 = \Delta = 0.2$. Then $\bar{M}(\tau^*, \Lambda) = 0.15 \times 10^{-3}$, $\bar{M}(\tau_1, \Lambda) = 0.50 \times 10^{-4}$, $\bar{M}(\tau_1, \Lambda) = 0.23 \times 10^{-5}$.

Thus, the use of the designs with uniformly distant points leads to a significant loss of accuracy in estimation.

REFERENCES

1. Adam, N.R.: Achieving a confidence interval for parameters estimated by simulation, Manage. Sci. 29 (1983), 856-866.

2. Alekseev, V.G.: On a technique for investigation of pseudo-random sequences, Zavodskaya Lab. 56 (3) (1990), 84-86 (in Russian).

3. Ahrens, J.H. and Dieter, V.: Non-uniform random numbers, Publisher, Graz, 1974, 356 pp.

4. Andrews, R.W. and Schriber, T.J.: Iteration analysis of output from PSS-based simulation, Proc. 1978 Winter Simulation Conf., New York, 1978, pp. 267-278.

5. Anisimov, V.V., Zakusilo, O.K. and Donchenko, V.S.: The elements of queuing theory and asymptotic system analysis, Kiev, 1987, 246 pp (in Russian).

6. Asmunssen, S.: Conjugate process and the simulation of ruin problems, Stochastic Process Appl. 20 (1985), 213-229.

7. Basharin, G.P., Bocharov, P.P. and Kogan, J.A.: Queuing analysis in computing nets, Moscow, 1989, 336 pp (in Russian).

8. Bellman, R.: Introduction to matrix analysis, New York, Toronto, London, 1960.

9. Billingsley, P.: Convergence of probability measures, New York, London, 1971.

10. Borovkov, A.A.: Probabilistic processes in queuing theory, Moscow, 1972, 358 pp (in Russian).

11. Borovkov, A.A.: Mathematical statistics. Parameters estimating, test of hypothesis, Moscow, 1984, 470 pp (in Russian).

12. Box, G.E.P. and Draper, N.R.: A basis for the selection of a response surface design, J. Amer. Statist. Assoc. 54 (1959), 622-654.

13. Bratley, P., Fox, B.L. and Schrage, L.E.: A guide to simulation, New York, 1987, 422 pp.

14. Buslenko, N.P.: Complex systems simulation, Moscow, 1978, 400 pp (in Russian).

15. Buslenko, N.P., Kalashnikov, V.V. and Kovalenko, I.N.: Lectures on the complex systems theory, Moscow, 1973. 440 pp (in Russian).

16. Cao, X.R.: Convergence of parameter sensitivity estimates in a stochastic experiment, IEEE Trans. Automat. Contr. 30 (9) (1985), 845-853.

17. Chentsov, N.N.: Pseudorandom numbers for Markov chain simulation, J. Vycheslit. Matem. i Matem. Fiziki 7 (3) (1967), 632-643 (in Russian).

18. Chentsov, N.N.: Statistics decisive laws and optimal outputs, Moscow, 1972. 520 pp (in Russian).

19. Chernoff, H.: Locally optimal designs for estimating parameters, Ann. Math. Statist. 24 (1953), 586-602.

20. Cox, D.R. and Smith, W.L.: Queues, London, 1961, 142 pp.

21. Crane, A.M. and Iglehart, D.L.: Simulating stable stochastic systems. I: General multiserver queues, J.A.S.M. 21 (1974), 103-113.

22. Crane, A.M. and Iglehart, D.L.: Simulating stable stochastic systems. II: Markov chains, J.A.S.M. 21 (1974), 114-123.

23. Dahl, W., Murhaug, B. and Newgord, G.: SIMULA-67 – the universal programming

language, 1967.

24. Deak, I.: Random number generators and simulation, Budapest, 1990, 342 pp.

25. De Vroye, L.: Non-uniform random variate generation, Berlin, 1986, 624 pp.

26. Dovgal, V.V.: Analysis of pseudo-random sequences distribution, Dissertation, Leningrad, 1989, 150 pp (in Russian).

27. Ermakov, S.M.: On the optimal unbiased designs of regression experiments Trudy Matem. Instituta Akad. Nauk 3 (1970), 252-258 pp (in Russian).

28. Ermakov, S.M.: Note on pseudo-random sequences, J. Vychislit. Matem. i Matem. Fiziki 12 (4) (1972), 1077-1082 (in Russian).

29. Ermakov, S.M.: Monte Carlo technique and the related problems, Moscow, 1975, 472 pp (in Russian).

30. Ermakov, S.M.: Random interpolations in the theory of experimental design, Comp. Stat. & Data Anal. 8 (1989), 75-80.

31. Ermakov, S.M., Brodsky, V.Z., Melas, V.B. and et al.: Mathematical theory of experimental design, Moscow, 1983, 392 pp (in Russian).

32. Ermakov, S.M. and Melas, V.B.: A duality theorem and iterative technique for h-optimal design, Vestnik Leningrad. Universiteta (3) (1982), 38-43 (in Russian).

33. Ermakov, S.M. and Melas, V.B.: On the path branching technique for the complex systems simulation, Izvestiya Vysshei Shkoly, Matematika (5) (1988), 11-16 (in Russian).

34. Ermakov, S.M. and Melas, V.B.: On optimal path branching for simulation of systems, determined by random processes, Izvestiya Akad.Nauk., Tekhnicheskaya Kibernetika (2) (1989), 64-69 (in Russian).

35. Ermakov, S.M. and Melas, V.B.: Mathematical experiment with complex stochastic system models, Sankt-Petersburg, 1993, 272 pp (in Russian).

36. Ermakov, S.M. and Mikhailov, G.A. Statistic simulation, Moscow, 1982, 294 pp (in Russian).

37. Ermakov, S.M., Pokhodzei, B.B. and Pavlov, A.V.: On a technique for multiple integrals evaluation with automatical step selection, J. Vychislit. Matem. and Matem. Fiziki 17 (3) (1977), 572-578 (in Russian).

38. Ermakov, S.M. and Zhyglavsky, A.A.: Mathematical theory of an optimal experiment, Moscow, 1987, 314 pp (in Russian).

39. Fedorov, V.V.: Optimal experimental design, New York, 1972.

40. Feller, W.: An introduction to probability theory and its applications, New York, Toronto (1) (1970), (2) (1971).

41. Fishman, G.S.: Accelerated accuracy in the simulation of Markov chains, Oper. Res. 31 (1983), 466-487.

42. Franclin, J.N.: Deterministic simulation of random processes, Math. Comp. 17 (1963), 28-59.

43. Gantmacher, F. R.: The theory of matrices, 4d ed., 1988, 552 pp. (in Russian).

44. Golstain, E.G.: Duality theory in mathematical programming and its applications, Moscow, 1971, 352 pp (in Russian).

45. Harris, T.: The theory of branching processes, Springer-Verlag, Berlin, 1963, 356

pp.
46. Jansson, B.: Random number generators, Stocholm, 1966.
47. Karlin, S. and Stadden, V.: Tchebysheff systems: with application in analysis and statistics, Wiley & Sons, New York, London, 1966.
48. Kac, M.: On the distribution of values of the type $f(2^k t)$, Ann. Math. 47 (1) (1946), 33-49.
49. Kahn, H.: Use of different Monte Carlo sampling techniques. Symposium on Monte Carlo methods, Wiley, New York, 1956, 146-190.
50. Kalashnikov, V.V. and Rachev, S.T.: Mathematical methods of stochastic queuing models construction, Moscow, 1988 (in Russian).
51. Kashtanov, Y.N.: Some ways to diminish the dispersion at evaluation of functionals of stationary Markov chain distribution by means of Monte Carlo technique, Vestnik Leningrad. Universiteta (7) (1981), 42-49 (in Russian).
52. Kuipers, L. and Niederreiter H. Uniform distribution for sequences, Wiley & Sons, New York, London, Sydney, Toronto, 1985.
53. Kelbert, M.Y. and Sukhov, Y.M.: Mathematical problems of the queuing network theory. In: Results of Sciences and Technology. Probability Theory. Mathematical Statistics. Cybernetics Theory. Moscow, 1988, 3-96 (in Russian).
54. Kemeny, J. and Snell, J.L.: Finite Markov chains, The University Ser. in Undergraduate Mathematics, 1960.
55. Kemeny, J., Snell, J.L. and Knapp, A.W.: Denumerable Markov chains, Springer-Verlag, 1976.
56. Kiefer, J.: General equivalence theory for optimum designs (approximative theory), Ann. Statist. 2 (1974), 849-879.
57. Kiefer, J. and Wolfowitz, J.: Optimum designs in regression problems, Ann. Math. Stat. 30 (1959), 271-294.
58. Kiefer, J. and Wolfowitz, J.: The equivalence to two extremum problems, Can. J. Math., 12 (1960), 363-366.
59. Khisamutdinov, A.I.: On the selection of "splitting-roulette"parameters for radiation transport calculation, J. Vychislit. Matem. and Matem. Fiziki 29 (2) (1989), 286-293 (in Russian).
60. Kleijnen, J.P.C.: Statistical techniques in simulation. Part I, New York, 1974, 205 pp.
61. Knuth, D.: The art of computer programming, 2, Seminumerical algorithms, Addison-Wesley Publ. Comp., 1969.
62. Knuth, D. and Yao, E.: The complexity of nonuniform random number generation. Algorithms and Complexity, Akad. Press., (1) (1976), 357-428.
63. Kolmogorov, A.M.: On tables of random numbers, Sankhya. Indian J. Stat. Serv. A. 25 (1963), 369-376.
64. Korobov, N.M.: On the completely uniform distribution of joint normal numbers, Izvestiya Akad. Nauk, Ser. Matematika. 20 (5) (1956), 649-660 (in Russian).
65. Korobov, N.M.: Theoretic numerical methods in approximate analysis, Moscow,

1963, 224 pp (in Russian).

66. Kovalenko, I.N.: Research on complex systems accuracy analysis, Kiev, 1976, 204 pp (in Russian).

67. Kovalenko, I.N.: The limit accuracy theorems, Kibernetica, (6) (1977), 106-116 (in Russian).

68. Kovalenko, I.N.: Rare events analysis when evaluating system efficiency and accuracy, Moscow, 1980, 208 pp (in Russian).

69. Kovalenko, I.N. and Kuznetsov, N.Y.: Calculation methods for systems with high accuracy, Moscow, 1988, 172 pp (in Russian).

70. Kovalenko, I.N. and Nakonetchny, A.N.: Approximate computations and optimization of accuracy, Kiev, 1989, 182 pp (in Russian).

71. Kubilus, I.P. and Linnik, Y.V.: Digital simulation of Brownian motion, Izvestiya Vysshei Shkoly (6) (1959), 88-89 (in Russian).

72. Levit, M.V.: On the selection of parameters for pseudo-random numbers generator, Izvestiya Akad. Nauk 37 (3) (1989), 529-533 (in Russian).

73. Lewis, P.W.A. and Oral, E.J.: Simulation methodology for statisticians, operations analysts and engineers, Belmont, 1989, 404 pp.

74. Linnik, I.Y.: The improvement of Monte Carlo technique convergence for some queuing problems, Kibernetica (5) (1973), 129-132 (in Russian).

75. Linnik, Y.V.: The least squares technique and foundations of observation processing theory, Moscow, 1962, 350 pp (in Russian).

76. Loeve, M.: Probability theory, Toronto, New York, London.

77. Malutov, M.B. and Fedorov, V.V.: On design of experiments, taking into account possible nonadequacity, Moscow, 1971, Preprint 18, 14 pp (in Russian).

78. Melas, V.B.: Optimal designs for exponential regression, Math. Operations forsch. Statist., Ser. Statistics 9 (1978), 45-59.

79. Melas, V.B.: A duality theorem and E-optimality, Zavodskaya Lab. (3) (1982), 48-50 (in Russian).

80. Melas, V.B.: On an approach to optimal design of simulation experiment, In: Statistic Methods and Models, Moscow, 1987, 77-86 (in Russian).

81. Melas, V.B.: On path branching method to increase efficiency of complex systems simulation, Kiev, 1990, Preprint 90-29, 22 pp (in Russian).

82. Melas, V.B.: On an approach to data compression for imitation experiment, Vestnik Leningrad. Universiteta, Ser. 1 (2) (1991), pp 40-47 (in Russian).

83. Melas, V.B.: Optimal simulation design by branching technique, In: Model-Oriented Data Analysis, (Eds. Muller, W.G., Wynn, H.P. and Zhigljavsky, A.A.), 1993, 113-126.

84. Melas, V.B.: Branching technique for Markov chain simulation (finite state case), Statistics 25 (1994), 159-171.

85. Melas, V.B.: Non-Tchebysheff E-optimal design and representation of positive polynomials, I, II, III, Vestnik St-Petersburg. Universiteta (1995), to be published (in Russian).

86. Mikhailov, G.A.: Weight Monte Carlo techniques optimization, Moscow, 1987, 234

pp (in Russian).
87. Mysovskih, I.P.: Imitational cubature formulas, Moscow, 1981, 336 pp (in Russian).
88. Niderreiter, H.: Random number generation and Quasi-Monte Carlo methods, Soc. for Ind. and Appl. Mathematics, Philadelphia, Pennsylvania, 1992.
89. Numellin, E.: General irreducible Markov chains and non-negative operators, London, New York, 1986.
90. Ogibin, V.N.: On applications of splitting and roulette in the calculation of particles transport by Monte Carlo technique, In: Monte Carlo technique in radiation transport problems, Moscow, 1967, pp 72-82 (in Russian).
91. Pokhodzei, B.B.: Combination of rejection technique with composition technique for continuous distributions modelling: "pseudoinversion" technique, Zavodskaya Lab. (10) (1988), 82-86 (in Russian).
92. Pukelsheim, F.: On linear regression designs which maximize information, J. Staist. Planning and Inference 4 (1980), pp 339-364.
93. Rao, S.R.: Linear statistical inference and its applications, 2nd ed., Wiley, New York, 1973.
94. Rasputin, G.G.: Cubature formulas and orthogonal polynomials common zeros, Vestnik Leningrad. Universiteta 3 (3) (1978), pp 40-45 (in Russian).
95. Sedunov, E.V and Ermakov, S.M.: Unbiased design and regression experiments analysis in a finite-dimensional function space, Vestnik Leningrad. Universiteta (1) (1974), 11-21 (in Russian).
96. Szego, G.: Orthogonal polynomials., revised edition, Am. Math. Soc. Coll. Pub. 23 (1959), New York.
97. Siegmund, D.: Importance sampling in the Monte Carlo study of sequential tests, Ann. Stat. 4 (1976), 673-684.
98. Sigman, K.: Queues as Harris recurrent Markov chain, Queuing system 3 (1988), 179-198.
99. Sobol, I.M.: Digital Monte Carlo techniques, Moscow, 1973, 312 pp (in Russian).
100. Sobol, I.M.: Multivariate quadrature formulas and Haar functions, Moscow, 1969, 288 pp (in Russian).
101. Tawsworth, R.C.: Random numbers generated by linear recurrence modulo two, Math. Comp. 19 (90) (1965). 201-209.
102. Uspensky, V.A. and Semionov, A.L.: Algorithm theory: the main discoveries and applications, Moscow, 1987, 288 pp (in Russian).
103. Weyl, H.: Uber die Gleichverteilung von Zahlen mod Eins, Math. Ann. 77 (1916), 313-352.
104. Wiese, W.L: Atomic transition probabilities, Nucl. Inst. and Methods 90 (1970), 25.
105. Wilson, J.R.: Variance reduction techniques for digital simulation, Amer. Journal of Mathematical and Management Sciences 4 (1984), 277-312.
106. Yanovitch, L.A.: Approximate calculation of functional integrals with respect to Gauss measures, Minsk, 1976, 38 pp (in Russian).

SUBJECT INDEX

Other *Mathematics and Its Applications* titles of interest:

W.L. Miranker: *Numerical Methods for Stiff Equations and Singular Perturbation Problems*. 1980, 220 pp. ISBN 90-277-1107-0

K. Rektorys: *The Method of Discretization in Time and Partial Differential Equations*. 1982, 470 pp. ISBN 90-277-1342-1

L. Ixary: *Numerical Methods and Differential Equations and Applications*. 1984, 360 pp. ISBN 90-277-1597-1

B.S. Razumikhin: *Physical Models and Equilibrium Methods in Programming and Economics*. 1984, 368 pp. ISBN 90-277-1644-7

A. Marciniak: *Numerical Solutions of the N-Body Problem*. 1985, 256 pp. ISBN 90-277-2058-4

Y. Cherruault: *Mathematical Modelling in Biomedicine*. 1986, 276 pp. ISBN 90-277-2149-1

C. Cuvelier, A. Segal and A.A. van Steenhoven: *Finite Element Methods and Navier-Stokes Equations*. 1986, 500 pp. ISBN 90-277-2148-3

A. Cuyt (ed.): *Nonlinear Numerical Methods and Rational Approximation*. 1988, 480 pp. ISBN 90-277-2669-8

L. Keviczky, M. Hilger and J. Kolostori: *Mathematics and Control Engineering of Grinding Technology. Ball Mill Grinding*. 1989, 188 pp. ISBN 0-7923-0051-3

N. Bakhvalov and G. Panasenko: *Homogenisation: Averaging Processes in Periodic Media. Mathematical Problems in the Mechanics of Composite Materials*. 1989, 404 pp. ISBN 0-7923-0049-1

R. Spigler (ed.): *Applied and Industrial Mathematics*. Venice-1, 1989. 1991, 388 pp. ISBN 0-7923-0521-3

C.A. Marinov and P. Neittaanmaki: *Mathematical Models in Electrical Circuits. Theory and Applications*. 1991, 160 pp. ISBN 0-7923-1155-8

Z. Zlatev: *Iterative Improvement of Direct Solutions of Large and Sparse Problems*. 1991, 328 pp. ISBN 0-7923-1154-X

M. Vajtersic: *Algorithms for Elliptic Problems*. 1992, 352 pp. ISBN 0-7923-1918-4

V. Kolmanovskii and A. Myshkis: *Applied Theory of Functional Differential Equations*. 1992, 223 pp. ISBN 0-7923-2013-1

A.D. Egorov, P.I. Sobolevsky and L.A. Yanovich: *Functional Integrals: Approximate Evaluation and Applications*. 1993, 426 pp. ISBN 0-7923-2193-6

G.I. Marchuk: *Adjoint Equations and Analysis of Complex Systems*. 1994 ISBN 0-7923-3013-7

Other *Mathematics and Its Applications* titles of interest:

A. Bakushinsky and A. Goncharsky: *Ill-Posed Problems: Theory and Applications.* 1994, 256 pp. ISBN 0-7923-3073-0

G.N. Milstein: *Numerical Integration of Stochastic Differential Equations.* 1995, 170 pp. ISBN 0-7923-3213-X

V.V. Shaidurov: *Multigrid Methods for Finite Elements.* 1995, 331 pp. ISBN 0-7923-3290-3

A.N. Tikhonov, A.V. Goncharsky, V.V. Stepanov and A.G. Yagola: *Numerical Methods for the Solution of Ill-Posed Problems.* 1995, 264 pp. ISBN 0-7923-3583-X

Batch number: 10583998

Printed by Printforce, the Netherlands